FLOWERS OF THE PACIFIC ISLAND SEASHORE

A guide

to the littoral plants of

HAWAI'I TAHITI SAMOA

TONGA COOK ISLANDS FIJI

and

MICRONESIA

by

DR. W. ARTHUR WHISTLER

BOTANIST, NATIONAL TROPICAL BOTANICAL GARDEN

1992

Published by Isle Botanica
500 University Avenue # 1601
Honolulu, Hawaii 96826 U. S. A.

Distributed by University of Hawaii Press
2840 Kolowalu St.
Honolulu, Hawaii 96822 U.S.A.

Printed by Everbest Printing Co. Ltd., Hong Kong

ISBN: 0-8248-1528-9

This book is dedicated to Anne Caprio, whose friendship, sensitivity, and enthusiasm have been such a positive influence on my life.

All photos in the text were taken by the author.

COVER: The beach morning-glory, *Ipomoea pes-caprae,* on an isolated beach on 'Upolu, Western Samoa.

FOREWORD

The dream of becoming a beachcomber on some distant tropical shore has always held a particular fascination for Westerners. To be able to spend one's day relaxing under a tropical sun and walking a palm-covered beach bordering a blue lagoon has seemed like the ideal escape from the everyday pressures of the hurried lifestyle of our Western world. Ever since its European discovery, Polynesia has captivated the imagination of adventurers looking for an island paradise. Now, with the advent of the age of jet travel, a visit to the islands is no longer an impossible dream. Every year thousands of people fly to the South Pacific to spend their vacation in the tropical sun.

One of the most remarkable features of Polynesia is the vast array of exotic flowers to be found growing everywhere in profusion. For those who are interested in nature, the easiest way to identify these exotic species is by using an illustrated book on tropical flowers. There are already a number of such books available, but most of these deal only with cultivated ornamental plants such as hibiscus, frangipani, and jasmine. These are well-suited to those who do not go beyond the confines of their hotel or local tourist attractions. But it is for those who seek out nature, away from the hotels, away from the tourist attractions, or even away from civilization, that this book is intended.

PREFACE

The basis for the present book is an earlier one, *Coastal flowers of the tropical Pacific*, published by the Pacific Tropical Botanical Garden (now known as the National Tropical Botanical Garden) in 1980. It featured 67 species, but since that time I have continued studies on the flora of Polynesia, Polynesian plant names, and the local uses for plants. This new book includes a total of 120 species, nearly double the number in *Coastal flowers of the tropical Pacific*. Most of the added species are from Hawai'i and the Society Islands, but some rare Cook Island, Tongan, and Samoan species, which I had not previously seen or photographed, have been included as well. With these additions, nearly every species in the native littoral flora of Polynesia, and most of those of Micronesia and Melanesia as well, are now covered.

More than just the inclusion of new species was needed for this edition: because of recent research by myself and other botanists during the last decade, much of the information in the original edition had become outdated and in need of revision. Some of the alterations involve scientific names that have been changed or corrected by recent research. Other alterations are in some Polynesian names, which in the original edition were based on erroneous information in the literature. Still other alterations are additions to the distribution records based on recent floristic studies in the Pacific islands. In addition to all these changes, most of the photographs from the original edition have been replaced by more recent and better ones.

Although this book is based on Polynesia, it is also well suited for the study of the littoral flora of Micronesia: 88 of the species covered here are found in Micronesia—a total higher than that of any Polynesian archipelago (Tonga has 87). With these alterations and additions, I believe that the present work can be used as a comprehensive guide by laymen and scientists alike to learn the native plants of the tropical Pacific shores. Hopefully, this may help to foster an awareness of the fragile condition of tropical floras, and highlight the need for measures to promote the protection of endangered species and areas of natural vegetation on tropical Pacific shores, as well as in native rainforests throughout the tropics.

W. Arthur Whistler
July 1992

CONTENTS

THE TROPICAL PACIFIC ISLANDS 1
LITTORAL SPECIES 4
Table 1. Distribution of coastal flowers in the Pacific Islands 5
LITTORAL VEGETATION 9
Herbaceous Strand 11
Littoral Shrubland 13
Pandanus Scrub 13
Littoral forest 13
Mangrove forest 15
THE NAMES OF PLANTS 17
THE SPECIES 18
Trees 19
Shrubs 48
Herbs 76
Vines 113
Grasses and Sedges 129
GLOSSSARY OF BOTANICAL TERMS 139
SELECTED BIBLIOGRAPHY 145
INDEX TO SCIENTIFIC NAMES 147
INDEX TO VERNACULAR NAMES 151

ABOUT THE AUTHOR

Art Whistler was born near Death Valley, California, to which he attributes his early love of plants and vegetation. After receiving a B.A. and an M.A. at the University of California, he served three years with the U.S. Peace Corps in Western Samoa where he taught high school biology. Resuming his schooling, he received a Ph.D. in Botany at the University of Hawai'i in 1979. Since then he has made numerous research trips to Samoa, Tonga, the Cook Islands, Tahiti, and elsewhere in the Pacific, working on the ethnobotany, medicinal plants, and flora of the islands. Currently he is a botanist with the National Tropical Botanical Garden, where he has been employed since 1984. He has published several books on the botany of the Pacific Islands, including *Polynesian herbal medicine* (1992), has written numerous scientific articles on medicinal plants, ethnobotany, and floristics of Polynesia, and is an adjunct Associate Professor with the Botany Department of the University of Hawai'i.

Photo by Paddy Ryan

Fig. 1. Tafahi, a "high island" in Tonga.

THE TROPICAL PACIFIC ISLANDS

Nearly all of the islands of Polynesia are oceanic islands, that is, islands formed from volcanic domes arising from the depths of the Pacific Ocean, and have never had a land connection with continental land masses. They are of two basic types, "high islands" and "low islands." High islands (Fig. 1) are usually composed of basalt, and may extend up to thousands of meters in elevation. They usually have rich and varied floras, with the number of species present being dependent upon the island's size and elevation, its distance from other islands or continents, its age, and the effects of human disturbance.

Low islands, on the other hand, are usually formed from coral reefs, but even these reefs are situated on the remnants of volcanic islands whose tops have either eroded away or have submerged below sea level. These are also called coral islands, and if the reef is in the form of a ring enclosing a lagoon, the island is called an "atoll" (Fig. 2). Atolls are the most typical type of low island, and are composed of one to many sand islets called "motu," which lie in a chain on the reef around a central lagoon, like pearls in a necklace. The highest elevation of the sand islets is usually less than 4 m. Because of the small

Fig. 2. Rose Atoll, American Samoa.

size and low elevation of atolls, the number of species present on them is very low compared to that found on high islands, and a high percentage of them are littoral species.

Intermediate with these two types of islands is a "makatea" island (from the island of Makatea in the Tuamotus), literally "white rock" (Fig. 3) These are actually atolls or other coral islands that for some reason have been raised above sea level. Their rugged limestone surfaces, often very jagged and porous from the effects of weathering, may reach an elevation of over 100 m. Examples of makatea islands are the outside portion of Mangaia in the Cook Islands and Niue Island in western Polynesia. Other islands may not fit well into any of the three categories; Tongatapu (Tonga), for example, has a structure similar to that of a makatea island, but has subsequently been covered by a layer of ash from a nearby volcano.

The islands of Polynesia are scattered like nighttime stars across the vast expanse of the Pacific Ocean, the largest geographical feature on

Fig. 3. Jagged makatea rock (karst) on Tinian Island, Micronesia.

earth. The boundaries of Polynesia are delineated to the north by the Hawaiian Islands, to the east by Easter Island, to the south by New Zealand (which is well out of the tropics), and to the west by Samoa or the "Polynesian outliers" that extend sporadically into Melanesia and Micronesia. Between these are Tonga, Niue, Tokelau, and Tuvalu in western Polynesia, and the Cook Islands and "French Polynesia," which comprises the Society Islands (including Tahiti), the Austral Islands, the Tuamotus, and the Marquesas in eastern Polynesia. In addition, Fiji, which lies to the west of Samoa, is often included in Polynesia; although its people are Melanesian, their culture is Polynesian.

Micronesia lies north and northwest of Polynesia, and like Polynesia, covers a vast area of the Pacific Ocean. Four main archipelagoes comprise it, the Marianas (including Guam), the Carolines (including Yap and Pohnpei), the Marshalls, and Kiribati (treated here to include the adjacent island of Nauru).

One hundred and twenty plants are included in this book, representing nearly all of the littoral species found in Polynesia. The few missing littoral species have been excluded because they are closely related to other species included in the text (and are discussed there), or because they are rare and have not been seen or photographed by the author. Three species not seen by the author are *Dioclea wilsonii* (Fabaceae), which is rare in Hawai'i and the Society Islands, *Digitaria stenotaphrodes* (Poaceae) of the low islands of the Society Islands and the Tuamotus, and *Pongamia pinnata* (Fabaceae), which has been collected only twice in Polynesia (Samoa).

Of the 120 species, 88 are found in Micronesia (including related littoral species), 87 in Tonga, 81 in Fiji, 81 in French eastern Polynesia, 76 in Samoa, 72 in Hawai'i, and 71 in the Cook Islands. A majority of these have very wide distributions, the prominent exception being some of the littoral species found in Hawai'i. Twelve of the species are endemic to Hawai'i, i.e., they are found nowhere else in the world; no other archipelago has any endemic species, with the exception of the Marquesas, which has one, *Nicotiana fatuhivensis* (noted under *Nicotiana fragrans*). This high number of endemic littoral species in Hawai'i, and the relatively low total number of littoral species there compared to the numbers occurring in other smaller archipelagoes, is a product of the relative isolation of Hawai'i from other land areas. See Table 1 for the exact distributions of the 120 species.

LITTORAL SPECIES

The subject of this book is coastal plants – the native or sometimes naturalized species one is likely to encounter on undisturbed or partially disturbed tropical shores of Polynesia. Coastal plants are often referred to more precisely as "littoral" species (*litoris* = shore in Latin). The category is complicated by the fact that a few species that are littoral in one archepelago are sometimes inland species in another. Not all plants found near the coast are littoral; some may be inland species that have been planted on the shore, and others are weeds that occur in a wide variety of habitats. Like other botanical groupings, the "littoral plant" category is not a precise term, because it is part of an artificial system we apply to the natural world for our own convenience, and it may be interpreted differently by different authors. Consequently, some of the species selected for this book are

Table 1. Distribution of coastal flowers in the Pacific Islands.

X—Native species or ancient introduction. I— Introduced and naturalized in recent times. R— Absent, but a related littoral species present. C— Recently introduced and commonly cultivated. E— Endemic. French Polynesia figures do not include Wallis & Futuna.

	Haw.	Fr. Poly.	Cook Islands	Samoa	Tonga	Fiji	Micro-nesia
TREES (29)							
Acacia simplex	-	-	-	X	X	X	X
Barringtonia asiatica	-	X	X	X	X	X	X
Bruguiera gymnorrhiza	I	-	-	X	X	X	X
Calophyllum inophyllum	X	X	X	X	X	X	X
Casuarina equisetifolia	I	X	X	X	X	X	X
Cerbera manghas	-	X	-	X	X	X	X
Cerbera odollam	-	X	X	X	X	-	R
Cocos nucifera	X	X	X	X	X	X	X
Cordia subcordata	X	X	X	X	X	X	X
Erythrina fusca	-	-	-	X	X	X	X
Erythrina variegata	C	X	X	X	X	X	X
Excoecaria agallocha	-	-	-	-	X	X	X
Guettarda speciosa	-	X	X	X	X	X	X
Heritiera littoralis	-	-	-	R	X	X	X
Hernandia nymphaeifolia	-	X	X	X	X	X	X
Hibiscus tiliaceus	X	X	X	X	X	X	X
Lumnitzera littorea	-	-	-	-	X	X	X
Neisosperma oppositifolium	-	X	X	X	X	X	X
Pandanus tectorius	X	X	X	X	X	X	X
Pisonia grandis	X	X	X	X	X	X	X
Rhizophora mangle	I	-	-	X	X	X	-
Rhizophora stylosa	-	I	-	-	X	X	X
Schleinitzia insularum	-	X	X	-	X	X	X
Terminalia catappa	I	I	I	X	X	X	X
Terminalia samoensis	-	X	R	X	R	R	X
Thespesia populnea	X	X	X	X	X	X	X
Tournefortia argentea	I	X	X	X	X	X	X
Xylocarpus granatum	-	-	-	-	X	X	X
Xylocarpus moluccensis	-	-	-	X	X	X	X

	Haw.	Fr. Poly.	Cook Islands	Samoa	Tonga	Fiji	Micro-nesia
SHRUBS (28)							
Batis maritima	I	-	-	-	-	-	-
Bikkia tetrandra	-	-	-	-	X	X	X
Caesalpinia bonduc	X	X	X	X	X	X	X
Capparis cordifolia	R	X	X	X	X	X	X
Chenopodium oahuense	E	-	-	-	-	-	-
Clerodendrum inerme	-	-	-	X	X	X	X
Colubrina asiatica	X	X	X	X	X	X	X
Corchorus torresianus	-	X	X	-	X	X	X
Dendrolobium umbellatum	-	-	-	X	X	X	X
Eugenia reinwardtiana	X	X	X	X	X	X	X
Ficus scabra	-	-	-	X	X	X	-
Gossypium hirsutum	I	X	-	X	-	X	X
Lycium sandwicense	X	X	-	-	X	-	-
Myoporum sandwicense	X	R	X	-	-	-	R
Pemphis acidula	-	X	X	X	X	X	X
Phyllanthus societatis	-	X	X	-	-	-	R
Premna serratifolia	-	X	X	X	X	X	X
Scaevola coriacea	E	-	-	-	-	-	-
Scaevola taccada	X	X	X	X	X	X	X
Sesbania tomentosa	E	R	-	-	R	R	-
Sida fallax	X	X	-	-	-	-	X
Sophora tomentosa	-	X	X	X	X	X	X
Suriana maritima	-	X	X	X	X	X	X
Timonius polygamus	-	X	X	-	X	X	-
Vitex rotundifolia	X	-	-	-	-	-	-
Vitex trifolia	C	X	X	X	X	X	X
Ximenia americana	-	X	X	X	X	X	X
Xylosma orbiculatum	-	-	-	-	X	X	-
HERBS (37)							
Achyranthes splendens	E	R	R	R	R	R	R
Achyranthes velutina	-	X	X	X	-	-	-

	Haw.	Fr. Poly.	Cook Islands	Samoa	Tonga	Fiji	Micro-nesia
Atriplex semibaccata	I	-	-	-	-	-	-
Bacopa monnieri	X	X	-	-	-	-	X
Boerhavia glabrata	X	X	-	R	-	-	R
Boerhavia repens	X	X	X	X	X	X	X
Boerhavia tetrandra	-	X	X	X	-	-	X
Chamaesyce atoto	-	X	X	X	X	X	R
Chamaesyce degeneri	E	-	-	-	-	-	—
Cressa truxillensis	X	-	-	-	-	-	-
Gnaphalium sandwicensium	E	-	-	-	-	-	-
Haloragis prostrata	-	-	X	-	-	-	-
Hedyotis biflora	-	-	X	X	X	X	X
Hedyotis foetida	-	X	X	X	X	X	X
Hedyotis romanzoffiensis	-	X	X	X	-	-	-
Heliotropium anomalum	X	X	X	-	-	-	X
Heliotropium curassavicum	X	-	-	-	-	-	R
Lepidium bidentatum	X	X	X	-	-	-	X
Lipochaeta integrifolia	E	-	-	-	-	-	-
Lysimachia mauritiana	X	R	-	-	-	-	X
Nama sandwicensis	E	-	-	-	-	-	-
Nesogenes euphrasioides	-	X	X	-	-	-	-
Nicotiana fragrans	-	R	-	-	X	-	-
Portulaca lutea	X	X	X	X	X	X	X
Portulaca samoensis	-	-	-	X	X	X	X
Portulaca villosa	E	-	-	-	-	-	-
Schiedea globosa	E	-	-	-	-	-	-
Sesuvium portulacastrum	X	X	X	X	X	X	X
Solanum amicorum	-	-	-	-	X	-	-
Solanum nelsonii	E	-	-	-	-	-	-
Tacca leontopetaloides	X	X	X	X	X	X	X
Tephrosia purpurea	X	X	X	X	X	X	R
Tetragonia tetragonioides	I	X	-	-	X	-	-
Tetramolopium rockii	X	-	R	-	-	-	-
Tribulus cistoides	X	X	X	-	-	-	X
Triumfetta procumbens	-	X	X	X	X	X	X
Wollastonia biflora	-	X	X	X	X	X	X

	Haw.	Fr. Poly.	Cook Islands	Samoa	Tonga	Fiji	Micro-nesia
VINES (16)							
Abrus precatorius	I	X	X	X	X	X	X
Canavalia cathartica	I	X	X	X	X	X	X
Canavalia rosea	-	X	-	X	X	X	X
Canavalia sericea	I	X	X	X	X	X	X
Cassytha filiformis	X	X	X	X	X	X	X
Cuscuta sandwichiana	E	-	-	-	-	-	-
Dalbergia candenatensis	-	-	-	-	X	X	X
Derris trifoliata	-	X	-	X	X	X	X
Entada phaseoloides	X	X	X	X	X	X	X
Ipomoea imperati	X	-	-	-	-	-	-
Ipomoea littoralis	X	X	X	X	X	X	X
Ipomoea macrantha	I	X	X	X	X	X	X
Ipomoea pes-caprae	X	X	X	X	X	X	X
Jacquemontia ovalifolia	X	-	-	-	-	-	-
Mucuna gigantea	X	X	X	X	X	X	X
Vigna marina	X	X	X	X	X	X	X
GRASSES & SEDGES (10)							
Cenchrus calyculatus	R	X	X	X	X	X	X
Cyperus stoloniferus	-	-	-	X	X	-	-
Fimbristylis cymosa	X	X	X	X	X	X	X
Ischaemum byrone	X	X	X	R	R	-	-
Lepturus repens	X	X	X	X	X	X	X
Mariscus javanicus	X	X	X	X	X	X	X
Paspalum vaginatum	I	I	I	I	I	I	I
Sporobolus virginicus	X	-	-	-	X	X	X
Stenotaphrum micranthum	-	X	X	X	X	X	X
Thuarea involuta	-	X	X	X	X	X	X
TOTALS 120	**72**	**81**	**71**	**76**	**87**	**81**	**88**

marginal littoral species that other authors may or may not include in their lists or books of littoral plants.

Although littoral species have a variety of life forms (e.g., tree, shrub, vine), they have a number of important characteristics in common. Most littoral species have buoyant, saltwater-resistant seeds that may be carried for long distances by sea currents. Most of those lacking this characteristic have instead sticky seeds that adhere to seabird feathers, allowing for long-distance dispersal to new littoral habitats. A few have fruits which may be eaten and transported internally by sea birds or migratory birds. These dispersal characteristics account for the extremely wide distributions for most littoral species. While nearly all of the species pictured here are "indigenous" (native to the region), very few of them are "endemic" to one island or archipelago. Some littoral plants are closely related to one or more inland species. Examples of this are *Scaevola taccada,* with several related inland species in Polynesia, and *Calophyllum inophyllum*, with a related inland species in western Polynesia and Fiji. The inland species that have evolved from littoral species often have different means of seed dispersal – from green fruits dispersed by seawater to red or purple fruits eaten by birds. This is an adaptation to dispersal in their new habitats.

Littoral plants are also tolerant of, or resistant to, salt spray, brackish ground-water, and even to occasional, although not prolonged, seawater inundation. Most of them also require bright light conditions for establishment and growth, a need which generally excludes them from shady forest habitats away from the shore. The methods of dispersal and the physiological characteristics that littoral plants share account for their typical restriction to a narrow zone of vegetation on the shore – they are limited inland by competition from the more vigorous species of the lowland and tropical rainforest, and seaward by the ocean. There are a few littoral plants, however, such as *Casuarina equisetifolia*, which may also thrive in inland habitats.

LITTORAL VEGETATION

Littoral species occur in and comprise several more or less distinct types of vegetation (communities) on tropical islands. Littoral communites differ from inland communities in both their area and distribution – they occupy a narrow area on the coast and typically exhibit zonation into several bands that run roughly parallel to the coastline. A case can be made for combining all of these zones into a single

Fig. 4. Sandy herbaceous strand on Moloka'i, Hawai'i.

Fig. 5. Rocky herbaceous strand on Savai'i, Western Samoa.

littoral or strand community: distinct boundaries between zones are often missing; some coasts lack some of the zones or communities; and their overall size is so small. However, since the zones are sometimes distinct, and they are often recognized in the literature, they will be treated here as distinct units.

Although names for the vegetation types vary somewhat in the literature, five main littoral plant communities may be distinguished: (1) herbaceous strand; (2) littoral shrubland; (3) pandanus scrub; (4) littoral forest; and (5) mangrove forest. The five communities are discussed below.

Herbaceous Strand

This narrow zone of vegetation occupies the upper portion of sandy or rocky beaches, limited inland by littoral forest or littoral shrubland, and seaward by the hightide mark of the ocean. It is sometimes divided into two subtypes – rock strand and sand strand – depending upon whether the substrate is rocky or sandy. However, since the species comprising the two types are often the same, and intermediates are common, they are combined here into one community, herbaceous strand.

Herbaceous strand on sand is particularly common on atolls, where nearly all the land surface is formed by sand islets called "motus." The dominant species are herbaceous creeping vines (Fig. 4), most typically, *Ipomoea pes-caprae* (beach morning-glory), *Vigna marina* (beach pea), and *Canavalia rosea*. Grasses, such as *Thuarea involuta* and *Lepturus repens,* are also frequently found here.

Herbaceous strand on rock substrate occurs on basalt, as on volcanic islands, or on limestone, as on makatea islands and coasts formed from upraised reefs. The habitat is very harsh, and only the hardiest of plants can survive in the depressions and cracks of the rock surface (Fig. 5). The salt spray, dry conditions, lack of soil, high temperatures, and occasional inundation by high waves discourage all but a few species from growing in this zone. The herbaceous vegetation on rock substrates is dominated by grasses and sedges (such as *Lepturus repens, Cyperus stoloniferus, Paspalum vaginatum*, and *Fimbristylis cymosa*), prostrate shrubs (such as *Pemphis acidula*), herbs (such as *Portulaca lutea* and *Sesuvium portulastrum*), or vines (such as those listed above for sandy substrates). Differences in species composition in rock strand vegetation are often related to differences in substrate –

Fig. 6. Littoral shrubland on 'Aunu'u, American Samoa.

Fig 7. Pandanus scrub thicket on Tutuila, American Samoa.

species occurring on or dominating coral rock substrate differ from those on basalt or lava rock.

Littoral Shrubland

This is the shrubby vegetation occurring on windy coastal ridges and slopes, as well as on the seaward margins of the littoral forest (Fig. 6). It is dominated by shrubby species up to 2 m or more in height, but these are sometimes prostrate or dwarfed by the action of strong, salty sea winds. Sometimes the boundary between this and the littoral forest is quite distinct, but often the two intergrade into each other. Two of the most characteristic species of littoral shrubland are *Scaevola taccada* and *Wollastonia biflora*.

Pandanus Scrub

This is the scrubby vegetation dominated by *Pandanus tectorius* (screwpine) typically occurring on rocky shores (Fig. 7). The sharp prickles on the trunks and leaves of *Pandanus* make passage through this type of vegetation difficult or painful. Where this vegetation occurs on exposed shores, it is often low and windswept, but in protected areas it may actually be a low forest dominated by *Pandanus,* which typically excludes all other tree species. The shade from the *Pandanus* trees and the accumulation of fallen leaves also prevents most herbaceous species from becoming established here. This community or zone is often absent on coasts, but when present, it may intergrade into the herbaceous strand on its seaward side, and the littoral forest on its inland side.

Littoral Forest

This is the most common and characteristic type of vegetation occurring on tropical shores. It consists of dense forest and is often dominated by a single tree species (Fig. 8). The most characteristic trees are *Barringtonia asiatica*, *Pisonia grandis*, *Hernandia nymphaeifolia*, *Casuarina equisetifolia*, *Guettarda speciosa*, and *Terminalia* spp. Although coconuts are common and sometimes dominant on Polynesian shores, they occur mostly in or near villages and coastal plantations, in areas where they have either been planted or are remnants of former cultivation. They rarely occur in undisturbed

Fig. 8. *Pisonia* littoral forest on Nu'ulua Islet, Western Samoa.

littoral forests because they are usually unable to compete with other more successful tree species in this habitat, especially on rocky shores.

The floor of littoral forests is typically open, because most littoral herbs and shrubs are "heliophytes" that require bright sunlight for germination and growth. Consequently, the ground cover is often dominated by ferns such as the bird's-nest fern (*Asplenium nidus*) and seedlings of the littoral trees. Epiphytes and vines are also uncommon in littoral forest, probably because of the harsh climatic conditions caused by the proximity of the sea.

Fig. 9. Mangrove forest at Sataoa-Sa'anapu, 'Upolu, Western Samoa.

Mangrove Forest

Mangrove forest (or mangrove swamp as it is sometimes called) is not "littoral" in the usual sense of the word. Instead of occupying narrow zones on the beach, it usually occurs on muddy, reef-protected areas that are periodically inundated by sea water. Mangrove forests (Fig. 9) are not naturally found in the Pacific east of Samoa—not because of lack of suitable habitat, but because of the absence of the appropriate species in these areas. There are a few patches of mangrove in the Society Islands and larger areas in Hawai'i, but these are dominated by species introduced in modern times.

Most mangrove species have specialized adaptations that enable them to exist in this harsh habitat. They are tolerant of, and may even require, sea water for growth. Most species have specialized roots; some have knobby breathing roots ("pneumatophores") that protrude above the muddy forest floor, or "prop roots" that spread from the trunk to assist them in obtaining oxygen for their underground roots. Additionally, most of the dominant species have seeds that germinate while still on the plant; these produce a long root before dropping to the ground as a developing seedling. In Polynesia, the most charac-

Fig. 10. Mangrove scrub at Nu'uuli, Tutuila, American Samoa.

teristic tree of mangrove forest is *Bruguiera gymnorrhiza*, a large tree with a spreading canopy that precludes all but a few other species from coexisting with it.

Sometimes a separate type of mangrove community is distinguished—mangrove scrub (Fig. 10) that is composed of shrubby mangroves rather than actual trees. This community is typically dominated by *Rhizophora mangle*, which is somewhat smaller than *Bruguiera* and is found in sunny localities such as estuary margins and even on sandbars in lagoons. It may later be displaced by the larger *Bruguiera*, as the vegetation matures and goes from scrub to forest. Several other mangrove species not found in Polynesia are common in Micronesia (see discussion under *Rhizophora stylosa*).

THE NAMES OF THE PLANTS

Several problems develop in trying to assign the correct common name to a plant species. It is not practical to use English names for these plants, since most of the native coastal species of the Pacific Islands lack such names. Using the Polynesian and Micronesian names also has drawbacks, since the plants are often known by similar but different names (i.e., "cognates") in the different archipelagoes (or even in the same one!), and some local names refer to more than one species of plant. The Polynesian and Micronesian names included in this book are taken from the sources listed in the bibliography, as well as from information gathered by the author during field work in the islands. Names not verified by the author include those of Fiji (Smith 1979–1991) and Guam (Stone 1970). The spellings of the local names are made in accordance with the most recent dictionaries of the different languages, and consequently are not always identical to those in the botanical literature. This is particularly the case for Fiji and Guam.

The most accurate way to identify the species is to use the Latin (scientific) name, but even this has its drawbacks. There are often several Latin names for the same species, and while all but one are technically incorrect, often only a botanist can know for sure which is the right one. The incorrect names are called synonyms; most of the pertinent ones have been recorded here under the appropriate species. A complete list of the valid names and synonyms can be found in the index at the back of the book—the former in regular type, the latter in italic.

The Polynsian languages are strictly phonetic in that each letter represents a single sound. The sounds are basically the same as those in European laguages, the most difficult one being the "ng" sound (spelled "g" in some Polynesian languages and "ng" in others) pronounced as in the word "song." There are never two consonants in a row, and all vowels are pronounced, even where there are three or more in a row. An apostrophe (glottal stop) in or before a word indicates there is a break between the vowel sounds on either side, and usually denotes that a consonant used in related languages has been replaced (cf. *'atae* of Tahiti and *ngatae* of Tonga). The accent on most Polynesian words is on the next to the last syllable (or vowel), except where there is a macron over another vowel. Thus *'avasā* (the Samoan name for *Tephrosia purpurea*), with the macron on the final vowel, is pronounced "ava-SA" rather than "a-VA-sa."

The Micronesian languages, including Chamorro, are distantly related to those of Polynesia, but many of the rules of grammar, such as no two consonants in a row, are different.

THE SPECIES

The plants on the following pages are arranged into five groups based upon their life form: (1) trees; (2) shrubs; (3) herbs; (4) vines; and (5) grasses and sedges. Within each group the plants are arranged in alphabetical order by Latin name. Some plants fit into more than one category, such as *Caesalpinia bonduc* (which is sometimes a shrub, but can also be found climbing into trees), so for these intermediate species a category was arbitrarily selected.

For each of the 120 species featured, the scientific name, English name (if any), and common names in Polynesia, Fiji, and Guam are listed. The subsequent text discusses the geographical range of the species, the habitats in which it grows, and uses of the plant throughout the Pacific Islands. Beside the photograph is a brief description of the species including synonyms (scientific names no longer considered to be correct) that have been used for the plant in earlier sources. A glossary is included at the back of the book to aid with unfamiliar botanical terms.

TREES

ACACIA SIMPLEX
Fabaceae (Pea family)

Vernacular names: *tatangia*—Tonga; *tatagia*—Samoa, Fiji

Small to medium-sized tree up to 10 m in height. Leaves (phyllodes) simple, alternate, blade elliptic to nearly round, mostly 8–16 cm long, base attenuate to a short petiole, surfaces glabrous, with 5–14 prominent, similar parallel veins. Flowers 25–30, in globose heads about 1 cm in diameter, on a peduncle 5–13 mm long. Calyx tiny, with 4–5 narrow lobes. Corolla of 4–5 tiny yellow petals. Stamens numerous, yellow, showy, exserted. Ovary superior. Fruit a narrow, oblong, flattened pod 6–15 cm long and 8–12 mm wide, somewhat constricted between the 3–10 dark, oval seeds 5–7 mm long. SYNONYMS: *Acacia simplicifolia, Acacia laurifolia.*

Acacia simplex is distributed from New Caledonia to western Polynesia (Fiji, Samoa, Tonga, Wallis Island, Horne Islands) and the Marianas (Rota). It is occasional to common in littoral forest on rocky and sandy coasts over most of its range; in Samoa, it is uncommon and restricted mostly to the western end of the island of Savai'i.

The wood is used for house timbers, posts, and carved handicrafts in Tonga, and reportedly for timber and ax handles in Fiji. A related endemic Hawaiian species, *Acacia koa,* is a dominant tree of inland forests; it is highly valued for its fine wood used commercially for making furniture and handicrafts.

BARRINGTONIA ASIATICA
Barringtoniaceae (Barringtonia family)

English name: fish-poison tree.
Vernacular names: *hutu* – Societies; *'utu* – Cooks; *futu* – Niue, Samoa, Tonga; *vutu* – Fiji; *puteng* – Guam

Large tree up to 20 m in height, often with a massive trunk and thick, spreading branches. Leaves simple, alternate, crowded at the branch tips, blade coriaceous, obovate to oblanceolate, 10 – 60 cm long, subsessile, surfaces glossy, glabrous. Flowers several, in short, terminal racemes. Calyx of 2 ovate sepals 3 – 4 cm long, pedicel 2 – 8 cm long. Corolla of 4 white, ovate petals 6 – 10 cm long. Stamens numerous, filaments white and pink, united at the base. Ovary inferior. Fruit large, ovoid, 4-angled, 8 – 12 cm long, husk fibrous, surrounding a single large seed. SYNONYM: *Barringtonia speciosa*.

Barringtonia asiatica is distributed from Madagascar to eastern Polynesia, and is found on most of the high islands of Polynesia (except Hawai'i, where it is rare in cultivation) and all the major archipelagoes of Micronesia. Although native in the western part of its range, it may be an ancient introduction in the east. It grows on rocky shores, where it often dominates the littoral forest, but is uncommon on atolls, perhaps being introduced to those on which it does occur.

The wood of the tree is not very durable, but is used for firewood and some construction. The most important part of the tree is its seed, which, when grated and spread in a lagoon, stuns fish, causing them to float to the surface where they are collected. The bark is used similarly in Tonga. Although the seed (and bark) is poisonous to fish, humans are unaffected by it. In the Cook Islands, the grated seed is used to treat burns, and the whole dry fruits have been widely used in Polynesia as fish-net floats.

BRUGUIERA GYMNORRHIZA
Rhizophoraceae (Mangrove family)

English name: oriental mangrove
Vernacular names: *togo* – *Samoa; tongolei* – Tonga; *dogo* – Fiji; *mangle machu* – Guam

Large tree up to 15 m or more in height, with large, red, stipules, a buttressed, fissured trunk, and spreading, knobby breathing roots. Leaves simple, opposite, blade elliptic, mostly 4 – 15 cm long, margins slightly revolute, surfaces glossy, glabrous, lower surface black-dotted. Flowers solitary, axillary, nodding. Calyx campanulate, 2.5 – 4 cm long, divided about halfway into 10 – 14 linear, red to yellow lobes. Corolla of 10 – 14 linear-lanceolate petals 12 – 15 mm long, brown, with 2 – 4 bristles at the tip. Stamens twice as many as petals, paired. Ovary inferior. Fruit top-shaped, 1 – 2.5 cm long, enclosed within the calyx, germinating on the plant to produce an angled, cigar-shaped root up to 15 – 25 cm long. SYNONYMS: *Bruguiera rheedii*, *Bruguiera conjugata* of some authors.

Bruguiera gymnorrhiza is distributed from East Africa to western Polynesia (Fiji, Tonga, Samoa) and all of the major archipelagoes of Micronesia, and is introduced and naturalized in Hawai'i. It grows in swampy coastal areas, often forming monodominant forests with an open floor and closed canopy (see Fig. 9).

Mangrove forest is useful in protecting the coast from hurricane damage, and provides an important breeding habitat for fishes. The wood of *Bruguiera gymnorrhiza* is sometimes used in construction, and in Tonga, a brown dye is obtained from the bark.

CALOPHYLLUM INOPHYLLUM
Clusiaceae (Mangosteen family)

English names: calophyllum, Alexandrian laurel
Vernacular names: *kamani*—Hawai'i; *'ati*— Societies; *tamanu*—Cooks; *fetau*—Niue, Samoa; *feta'u*—Tonga; *dilo*—Fiji; *da'ok*—Guam

Large tree up to 20 m in height, with 4-angled stems and yellow latex. Leaves simple, opposite, blade coriaceous, elliptic, 10–25 cm long, finely pinnately veined, surfaces glabrous, petioles 1–3 cm long. Flowers many, in axillary raceme-like inflorescences 4–15 cm long. Calyx of 4 orbicular sepals 4–8 mm long, the inner pair white inside. Corolla of 4 white, orbicular petals 8–14 mm long. Stamens numerous, yellow. Ovary superior. Fruit a globose drupe 2–3.5 cm long, green to yellow at maturity.

Calophyllum inophyllum is distributed from East Africa to Hawai'i, and is found on nearly all the high islands of Polynesia and Micronesia, but it is probably an ancient introduction in the eastern part of its range. It usually grows on rocky shores and sometimes forms almost pure stands in littoral forest, especially on cliff-bound coasts. It does not thrive on atolls, and is probably planted on those where it does occur.

The highly prized wood is widely used for making canoes, bowls, furniture, gongs, and houses in Polynesia. The oil extracted from the seed is mixed with coconut oil and used in massage and hair care. The tree is also commonly employed in native medicines: a solution of the crushed leaves is reportedly used as an eyewash and as a treatment for skin infections in the Society Islands, Cook Islands, Tonga, and Samoa.

CASUARINA EQUISITIFOLIA
Casuarinaceae (Casuarina family)

English names: ironwood, casuarina
Vernacular names: *'aito* – Societies; *toa* – Cooks, Niue, Samoa, Tonga; *nokonoko* – Fiji; *gagu* – Guam

Large tree up to 20 m in height, with slender, longitudinally striate, pineneedle-like branches drooping at the tips. Leaves reduced to minute scales in whorls of 6–9 around the nodes. Flowers many, in separate male and female inflorescences. Calyx and corolla absent. Male flowers in dense spikes 1–8 cm long, female flowers in short capitate spikes 2–5 mm across. Stamen 1. Ovary superior, with two red styles. Fruit a subglobose to oblong cone 1–2 cm long, composed of many 1-seeded nuts. SYNONYM: *Casuarina litorea.*

Casuarina equisetifolia is distributed from Southeast Asia eastward throughout all the major high archipelagoes of Micronesia and Polynesia. It may, however, have been an ancient introduction to the eastern part of its range, and is a modern introduction to Hawai'i. Ironwood is common on rocky coasts of many high islands (except, perhaps, on volcanic islands such as Samoa) and also on inland lowland ridges and fernlands, where it is often the dominant tree.

Although infrequent on atolls, it is often a dominant species on the shores of makatea islands, and is sometimes cultivated in villages. The wood is very hard, and the name *toa* is also the Polynesian word for warrior. Ironwood has been widely used to make weapons, canoe parts, house posts, tool handles, fish hooks, and many other artifacts throughout the Pacific islands. In the Society Islands, the sap was used to make a red-brown dye, and the young stems are occasionally used to treat internal ailments. In the Cook Islands and Tonga, an infusion of the bark is used to treat mouth infections.

CERBERA MANGHAS
Apocynaceae (Dogbane family)

Vernacular name: *leva* – Samoa, Tonga (Niuatoputapu)

Small to medium-sized tree up to 12 m in height, with thick, succulent stems and milky latex. Leaves simple, spirally arranged and crowded at branch tips, blade elliptic to oblanceolate, 9–20 cm long, margins entire, surfaces glabrous, glossy, petiole 1–4 cm long. Flowers several, in terminal cymes 10–30 cm long. Calyx split to near base into 5 white, oblong to oblanceolate sepals 1.5–3 cm long. Corolla salverform, white with a red throat, tube 2.5–4.5 cm long, lobes 5, oblong, 1–2 cm long. Stamens 5. Ovary superior. Fruit an ovoid drupe 5–10 cm long, red, fleshy on the surface, fibrous within. SYNONYM: *Cerbera lactaria.*

Cerbera manghas is distributed from the Seychelles in the Indian Ocean to eastern Polynesia. In Polynesia, it is found in the Marquesas, Pitcairn, Samoa, Tonga (Niuatoputapu and Niuafo'ou), the Horne Islands, and 'Uvea, and in the Carolines and Marshalls of Micronesia. In western Polynesia, it grows in littoral forest and in disturbed coastal areas, but in the Marquesas it (or perhaps what should be considered a different species or subspecies) ranges from the shore up into the mountains at over 1000 m elevation. A related species, *Cerbera odollam*, is found in scattered localities in Polynesia, and a third species, *Cerbera dilatata*, is an inland tree on Guam.

The milky latex is poisonous, and the fruit was reportedly eaten as a means of suicide in the Marquesas. Its occasional use in native medicines is reported from Samoa, Tonga, and elsewhere. The tree is sometimes planted around houses for its attractive flowers.

CERBERA ODOLLAM
Apocynaceae (Dogbane family)

Vernacular names: *reva* – Societies, Cooks; *leva* – Samoa; *toto* – Tonga

Medium-sized tree up to 10 m or more in height, with thick, succulent stems and milky latex. Leaves simple, spirally arranged and crowded at branch tips, blade elliptic to oblanceolate, 12–30 cm long, surfaces glabrous, glossy, petiole 2–5 cm long. Flowers several, in terminal cymes 8–30 cm long. Calyx split to near the base into 5 obovate to elliptic, pale green sepals 10–15 mm long. Corolla salverform, white with a yellow, hairy throat, tube 1.5–2.2 cm long, cup-shaped at top, lobes 5, obovate, 2–3 cm long. Stamens 5. Ovary superior. Fruit an ovoid drupe 5–10 cm long, red, fleshy on the surface, fibrous within. SYNONYM: *Cerbera manghas* of some authors.

Cerbera odollam is distributed from tropical Asia to eastern Polynesia, where it occurs in the Society Islands, Australs, Cooks, Tonga, and Samoa. It grows in littoral forest in Tonga, but elsewhere in Polynesia is more often found inland in forests up to 700 m elevation.

The tree is sometimes cultivated for its showy flowers, and is occasionally used in native medicines in the Society Islands, Austral Islands, Tonga, and elsewhere in Polynesia, but the fruit and sap are poisonous. It has often been confused with *Cerbera manghas* of western Polynesia and the Marquesas, but differs in having wider sepals and a shorter corolla tube that is yellow rather than red in the center.

COCOS NUCIFERA
Arecaceae (Palm family)

English name: coconut
Vernacular names: *niu* – Hawai'i, Niue, Samoa, Tonga, Fiji; *nū* – Cooks; *ha'ari* – Societies; *niyok* – Guam

Palm tree up to 20 m or more in height. Leaves spirally arranged, 6 m or more in length, pinnately divided into numerous strap-shaped segments. Flowers unisexual, arranged in 3s (2 male and 1 female, but only males are produced towards the tip) in axillary panicles up to 1 m or more in length, emerging from a woody, boat-shaped sheath. Calyx of 3 yellow, overlapping sepals. Corolla of 3 yellow petals. Stamens 6 in male flowers. Ovary in female flowers superior. Fruit large, ovoid, up to 20 – 30 cm long, 1-seeded, with a fibrous husk surrounding the hard shell, brown at maturity.

Cocos nucifera is probably native to the Old World, but is now found throughout the tropics. Although easily dispersed by seawater, it was probably an ancient introduction to some of its Polynesian range (including Hawai'i), and prior to the 15th century was restricted in the New World to the Pacific Coast of Central America. It grows on sandy shores and inland up to about 500 m elevation, often in plantations, but if untended, will usually be replaced by other tree species.

The coconut is the most valuable tree in the Pacific islands, with nearly every part of the plant having some use: the "meat" of the nut is used for food and for making coconut oil, and the "water" for drinking; the shell is used to make cups and charcoal; the husk is used to make sennit for ropes and cords, and as kindling; the sap from the cut flowering stalk is collected and naturally fermented into "toddy;" the leaves are used in weaving and plaiting, and are second in importance only to those of pandanus; and the trunk is used for timber and fashioned into various artifacts.

CORDIA SUBCORDATA
Boraginaceae (Borage family)

English name: cordia
Vernacular names: *kou*—Hawai'i; *tou*—Societies, Cooks; *motou*—Niue; *tauanave*—Samoa; *pua taukanave*—Tonga; *nawanawa*—Fiji; *niyoron*—Guam

Medium-sized tree up to 10 m in height. Leaves simple, alternate, blade broadly ovate to elliptic, 7–30 cm long, surfaces mostly glabrous, petiole mostly 2–8 mm long. Flowers in loose axillary or terminal cymes. Calyx urn-shaped, 10–17 mm long, 5–10-lobed. Corolla broadly trumpet-shaped, orange, wrinkled, 3–4.5 cm long, shallowly 5–7-lobed. Stamens as many as the corolla lobes. Ovary superior. Fruit a subglobose drupe 2–3.3 cm long, enclosed within the persistent calyx, drying black at maturity.

Cordia subcordata is distributed from tropical Asia to Hawai'i. Although found on all the major archipelagoes of Polynesia and Micronesia, it is probably an ancient introduction over much of the eastern part of its range, including Hawai'i. It grows in littoral forest and thickets on atolls and on sandy shores of high islands, but rarely very far inland.

The fine-grained, highly prized wood is used for making plank canoes, bowls, furniture, paddles, and drums. The showy flowers are used to make leis, and in some places in Polynesia the inner bark is fashioned into dancing skirts, mats, hats, baskets, and so forth. The small, edible seeds have been used as food in times of famine, and in the Society and Cook Islands, various parts of the plant are ingredients in native medicines used to treat internal ailments.

ERYTHRINA FUSCA
Fabaceae (Pea family)

Vernacular names: *ngatae fisi* – Tonga; *lalapā?, gatae pālagi* – Samoa; *drala kaka* – Fiji

Medium-sized tree 5–10 m in height, with spiny branches and often a buttressed trunk. Leaves alternate, trifoliate, rachis mostly 8–15 cm long, leaflet blades ovate to elliptic, 10–20 cm long, surfaces glabrous. Flowers many, in fascicles on terminal or axillary racemes mostly 10–30 cm long. Calyx 10–15 mm long, irregularly and shallowly lobed, often split on one side. Corolla papilionaceous, red to orange with white splotches, 3.5–5 cm long. Stamens 10, diadelphous. Ovary superior. Fruit a linear, slightly compressed pod 15–30 cm long, with 6–12 black to brown seeds 12–18 mm long.

Erythrina fusca is pantropical in distribution, but in Polynesia does not extend east of Tonga and Samoa, and in Micronesia is only found in the Carolines. It is apparently a Polynesian or modern introduction to Tonga (and probably Samoa), since its Tongan name means "Fijian coral tree." It grows in freshwater swamps and marshes, and sometimes along streams, but rarely very far inland.

No uses are reported for this tree. The wood is fairly soft and, like that of *Erythrina variegata*, is not suitable for construction.

ERYTHRINA VARIEGATA
Fabaceae (Pea family)

English name: coral tree
Vernacular names: *wiliwili haole* – Hawai'i; *'atae* – Societies; *ngate – Niue; ngatae* (*gatae*) – Cooks, Samoa, Tonga; *drala dina* – Fiji; *gaogao – Guam*

Spreading tree up to 20 m in height, with trunk and branches coarsely spiny. Leaves alternate, trifoliate, rachis 6–25 cm long, leaflet blades ovate to orbicular, 4–25 cm long, surfaces glabrous at maturity. Flowers many, in axillary racemes up to 35 cm or more long. Calyx narrow-ovoid, entire to slightly lobed, mostly 2–3 cm long. Corolla showy, papilionaceous, orange-red, 4–6 cm long. Stamens 10, diadelphous. Ovary superior. Fruit a curved, linear-oblong pod 12–22 cm long, containing 3–10 kidney-shaped seeds 10–15 mm long. SYNONYM: *Erythrina indica.*

Erythrina variegata (var. *orientalis*) is distributed from Zanzibar in the Indian Ocean to eastern Polynesia. It is found on most of the high archipelagoes of Polynesia and Micronesia (possibly as an Polynesian introduction on some islands), and is a modern introduction elsewhere in the tropics, including Hawai'i. It is most commonly found in littoral forest on rocky shores of high islands, and sometimes inland in coastal and ridge forests.

The wood is very light, and has little use other than for fish-net floats and firewood. Its flowering period, from July to September in the South Pacific, is sometimes used as a seasonal indicator. Two closely related endemic inland species are found in Polynesia, *Erythrina tahitensis* (now very rare in Tahiti) and *Erythrina sandwicensis* (common in Hawai'i).

EXCOECARIA AGALLOCHA
Euphorbiaceae (Spurge family)

Vernacular names: *fetānu* – Niue; *feta'anu* – Tonga; *sinu dina* – Fiji; *butabuta*? – Guam

Medium-sized tree up to 12 m or more in height, with a milky latex. Leaves simple, alternate, blade elliptic to obovate, 4–12 cm long on a short petiole, surfaces glabrous, glossy. Flowers several to many, in narrow, dense, axillary spikes or racemes 2–7 cm long, unisexual, plants dioecious. Calyx tiny, 3-lobed, sessile in male flowers, on a pedicel 1–2 mm long in female flowers. Corolla absent. Stamens of male flowers 2 or 3, exserted. Ovary of female flowers superior, 3-celled. Fruit a subglobose, 3-lobed schizocarp 3–5 mm long, splitting at maturity into three 1-seeded sections.

Excoecaria agallocha is distributed from India to Micronesia (Marianas, Carolines) and western Polynesia (Fiji, Tonga, Niue). It commonly grows on the seaward margin of littoral forest on limestone coasts, in coastal thickets, on coralline seacliffs, and in mangrove forests, but rarely very far inland.

The poisonous, milky latex causes intense irritation to the skin, and if rubbed into the eyes, can lead to blindness; even the smoke is extremely acrid and can cause severe eye irritation. The bark is occasionally used in medicines in Fiji, sometimes for treating the excruciating pain caused by the stings of poisonous fish.

GUETTARDA SPECIOSA
Rubiaceae (Coffee family)

Vernacular names: *tafano* – Societies; *'ano* – Cooks; *panopano* – Niue; *puapua* – Samoa; *puopua* – Tonga; *buabua* – Fiji; *panao* – Guam

Shrub or tree up to 10 m or more in height, with interpetiolar stipules. Leaves simple, opposite, blade obovate, 12–30 cm long, surfaces glabrous or pubescent on lower side. Flowers several, in scapose cymes 4–20 cm long, in upper axils. Calyx campanulate, 4–7 mm long, entire to shallowly lobed, on a short, thick pedicel. Corolla salverform, white, fragrant, velvety pubescent on the outside, tube 2.5–4 cm long, with 4–9 oblong to obovate lobes 6–10 mm long. Stamens epipetalous, as many as corolla lobes. Ovary inferior. Fruit an ovoid drupe 2–3 cm in diameter, brown at maturity.

Guettarda speciosa is distributed from East Africa to eastern Polynesia, and is found in all of the major Polynesian and Micronesian archipelagoes except Hawai'i. It grows on the margin and interior of littoral forests on rocky and sandy shores, and sometimes inland up to an elevation of 300 m.

Although the wood is of mediocre quality, it is very useful on atolls (where few tree species occur) for making house and boat parts, fishing poles, furniture, and other artifacts. The fragrant flowers are used in leis and are soaked in coconut oil to impart an aroma. The large leaves are often used to wrap food for cooking.

HERITIERA LITTORALIS
Sterculiaceae (Cacao family)

English name: looking-glass tree
Vernacular names: *mamea*—Tonga; *rosarosa*?—Fiji; *ufa*—Guam

Tree up to 10 m in height, with a buttressed trunk. Leaves simple, alternate, blade elliptic to ovate, 8–24 cm long, margins entire to wavy, upper surface glabrous, lower surface covered with silvery scales, petiole 5–20 mm long. Flowers many, in axillary panicles 2–6 cm long, unisexual, plants monoecious. Calyx campanulate, 3–6 mm long, shallowly divided into 4 or 5 ovate lobes, densely hairy on the outside, red inside. Corolla absent. Stamens 4–6, vestigial in female flowers. Ovary superior, composed of 4–6 separate carpels, vestigial in male flowers. Fruit composed of 2–5 irregularly ovoid nuts 4–7 cm long, with a flattened wing extending longitudinally around it.

Heritiera littoralis is distributed from East Africa eastward to Fiji and Tonga, and in Micronesia occurs only in the Marianas and Carolines. It grows on the edge of mangrove swamps, in coastal thickets, and in littoral forest, but rarely very far inland. Although reported to be locally abundant in Fiji, it is uncommon in Tonga.

No significant uses have been reported for this tree in Fiji or Tonga, but elsewhere the durable wood is widely used for boat building. A related species, *Heritiera ornithocephala*, is found in Fiji, Samoa, Tonga (Kao), and Niue, but differs in being an inland species with smaller leaves and longer petioles.

HERNANDIA NYMPHAEIFOLIA
Hernandiaceae (Hernandia family)

English name: Chinese-lantern tree
Vernacular names: *ti'anina* – Societies; *puka* – Cooks; *puka kula* – Niue; *pu'a* – Samoa; *fotulona* – Tonga; *evuevu* – Fiji; *nonak* – Guam

Large tree up to 20 m in height, with a massive trunk and branches. Leaves simple, alternate, blade peltate with a red or yellow spot on the upper surface, ovate, 7 – 20 cm long, surfaces glabrous. Flowers mostly in clusters of 3, in axillary cymes on a long peduncle mostly 8 – 20 cm long, unisexual, plants monoecious. Male flowers with 3 sepals, 3 white petals, and 3 stamens. Female flowers with 4 green sepals, 4 white petals, and a superior ovary. Fruit an ellipsoid drupe 15 – 23 mm long, surrounded by a globose vesicle open at the top, mostly 2.5 – 3.5 cm long, and yellow or red at maturity. SYNONYMS: *Hernandia peltata*, *Hernandia ovigera* of some authors, *Hernandia sonora* of some authors.

Hernandia nymphaeifolia is distributed from tropical East Africa to eastern Polynesia, and is found on all of the major archipelagoes of Micronesia and Polynesia except Hawai'i and the Marquesas. It is a common tree in littoral forest on sandy shores, often as a dominant species, but only rarely is it found very far inland. It is related to several endemic inland species in French Polynesia, and to *Hernandia moerenhoutiana*, a common inland tree in Polynesia.

The light wood is commonly used to make canoes and their outriggers. The round seeds are fashioned into leis and dancing skirts, as well as being used as marbles by children. In Tokelau, the leaves are used in native medicines to treat sore eyes.

HIBISCUS TILIACEUS
Malvaceae (Mallow family)

English name: beach hibiscus
Vernacular names: *hau* – Hawai'i; *purau* – Societies; *'au* – Cooks; *fou* – Niue; *fau* – Samoa, Tonga; *vau* – Fiji; *pago* – Guam

Shrub to tree up to 10 m or more in height, with paired lanceolate stipules 1–6 cm long. Leaves simple, alternate, blade cordate to orbicular, 8–30 cm long, upper surface green, lower surface gray and tomentose. Flowers axillary, solitary or in few-flowered cymes. Calyx cup-shaped, 1.5–3 cm long, divided into 10–12 lanceolate lobes with a cup-shaped ring of bracts below. Corolla of 5 obovate petals 4–6 cm long, lemon yellow with purple at the base. Stamens many, monadelphous. Ovary superior. Fruit an oblong to ovoid capsule 1.3–2.8 cm long, splitting open by 5 valves. SYNONYM: *Pariti tiliaceus*.

Hibiscus tiliaceus is a pantropical species found on nearly all the major archipelagoes of Polynesia and Micronesia. Its natural distribution is not certain, but the tree was possibly an ancient introduction to the eastern part of its Pacific range. It is a common or even dominant species of coastal thickets and streambanks, as well as in secondary forest of high islands, but is infrequent on atolls.

The beach hibiscus is one of the most useful of Pacific trees. Its light wood is used for timber, fish-net floats, firewood, and many other things. The inner bark fibers are fashioned into "grass skirts," cordage, mats, fishing line, and even reef shoes. The leaves are used for wrapping food (in cooking) and as serving platters. The slimy sap and other parts of the plant have been widely used in native medicines throughout Polynesia; in Tahiti, for example, the flowers were used to treat sores.

LUMNITZERA LITTOREA
Combretaceae (Tropical-almond family)

Vernacular names: *hangale* – Tonga; *sagali* – Fiji; *nganga, bakau-aine* – Guam

Shrub or small tree up to 9 m in height. Leaves simple, alternate, blade oblanceolate to elliptic, 5–10 cm long, sessile, margins slightly revolute, surfaces glabrous. Flowers several, in short, axillary racemes up to 5 cm long. Calyx 3–5 mm long, deeply divided into 5 lobes. Corolla of 5 bright-red, deciduous petals 4–6 mm long. Stamens 5–10, exserted. Ovary inferior. Fruit spindle-shaped with rounded angles, 1.4–3.2 cm long, with a persistent calyx on top, brown at maturity, 1-seeded.

Lumnitzera littorea is distributed from tropical Asia to Fiji and Tonga, and occurs on all the major archipelagoes of Micronesia. It is occasional to common along the edges of mangrove swamps and sometimes in open, dry places along the edges of littoral forest.

The timber is hard and durable, and is reportedly used in Fiji for pilings since it apparently resists marine borers. The showy red flowers are sometimes used in Tonga for making leis. Like other mangroves, it is important in providing a breeding habitat for inshore fishes.

NEISOSPERMA OPPOSITIFOLIUM
Apocynaceae (Dogbane family)

Vernacular names: *pao* – Niue; *fao* – Samoa, Tonga; *vao* – Fiji; *fago* – Guam

Small to medium-sized tree up to 10 m or more in height, with milky latex. Leaves simple, in whorls of 3 or 4, blade oblong to obovate, 10–30 cm long, upper surface glossy, petiole 1–5 cm long. Flowers several, in axillary or terminal cymes. Calyx cup-shaped, 2–3.5 mm long, deeply divided into 5 ovate sepals. Corolla sympetalous, salverform, white with yellow in the center, tube 4–6 mm long, lobes five, 7–11 mm long, curved back at maturity. Stamens 5. Ovaries 2, superior, sharing one stigma. Fruits paired, ellipsoid drupes 5–8 cm long, green, fibrous within, 1-seeded. SYNONYMS: *Ochrosia parviflora, Ochrosia oppositifolia.*

Neisosperma oppositifolium is distributed from the Seychelles in the Indian Ocean to Tahiti (where it may now be extinct). It is found in most of the archipelagoes in this area (except the Cooks), and is reported from all of the major archipelagoes of Micronesia except Kiribati (the Gilberts). It grows in littoral forest on sandy shores of high islands and atolls, but is not very common on the latter. A related inland endemic species of *Neisosperma* is found in the Marquesas.

The tree is of little reported use in Polynesia. The soft wood is used only for light construction, tool handles, and firewood. The wafer-like seed is edible, but is eaten only casually or in times of famine. In Tonga, the bark soaked in water is used as a potion in treating hypertension and diabetes.

PANDANUS TECTORIUS
Pandanaceae (Pandanus family)

English names: screwpine, pandanus
Vernacular names: *hala* – Hawai'i; *fara* – Societies; *'ara* – Cooks; *fā* – Niue, Tonga; *fala* – Samoa; *balawa* – Fiji; *kafo* – Guam

Shrub or tree up to 8 m or more in height, with prop roots and prickly stems. Leaves spirally arranged at branch tips, linear, up to 2 m or more in length, M-shaped in cross section, margins and midrib of lower surface prickly. Flowers unisexual, plants dioecious. Male flowers tiny, in dense, white inflorescences with large white, lanceolate bracts. Female flowers in a head. Fruit a subglobose, woody syncarp breaking up into woody phalanges with 1–several stigmas, husk fibrous, 1- to several-seeded. SYNONYMS: *Pandanus odoratissimus* of some authors, and many other names.

Pandanus tectorius is native to all the archipelagoes of Polynesia, and probably all of Micronesia as well, but its natural range is uncertain. Some botanists call it *Pandanus odoratissimus*; others divide it into numerous species based on minor differences in the shape of the phalanges, and treat many cultivated varieties as species. *Pandanus* is common on rocky and sandy shores of atolls and high islands, sometimes forming pure thickets on the coasts, and in some places is a dominant understory species in littoral forests.

The screwpine is one of the most useful trees in Polynesia, mainly for its leaves, which are woven or plaited into mats, thatch, sails, baskets, hats, and many other items. The fruit of cultivated varieties is a major source of food on atolls, and the prop roots can also be eaten. The tips of the prop roots are often used in native medicines in the Society Islands, Cook Islands, and elsewhere. The fragrant male flowers are used to scent coconut oil. The wood of the trunk is used in making native houses, and the split prop roots for house walls.

PISONIA GRANDIS
Nyctaginaceae (Four-o'clock family)

Vernacular names: *pu'a tea* – Societies; *puka tea* – Cooks; *puka sea* – Niue; *pu'a vai* – Samoa; *puko* – Tonga; *buka* – Fiji (Lau); *omumu* – Guam

Large tree up to 20 m in height, often with massive trunks. Leaves simple, opposite or appearing whorled, blade elliptic, mostly 7–25 cm long, surfaces mostly glabrous, petiole 1–4 cm long. Flowers many, in subumbellate clusters arranged in terminal cymes 3–10 cm long, unisexual, plants monoecious or dioecious. Calyx campanulate, 3–6 mm long, shallowly 4–6-lobed, green. Corolla absent. Stamens of male flowers 8–10. Ovary of female flowers inferior. Fruit narrowly cylindrical, 15–25 mm long, sticky and covered with short spines, 1-seeded, black at maturity.

Pisonia grandis is distributed from Madagascar to eastern Polynesia and Micronesia, and is found in all the major archipelagoes of the area. It is probably extinct, however, on the main islands of Hawai'i, where it has not been collected in over 150 years (a single tree was recently reported from the Leeward chain). It is most common on sandy shores of high islands and atolls, often being the dominant tree in littoral forest (see Fig. 8), and is occasionally found inland up to 300 m elevation (Marquesas). The sticky seeds are dispersed by adhering to seabird feathers.

The wood is soft and weak, and is used only in light construction. The edible young leaves are cooked and eaten as a vegetable in the Tuamotus, and elsewhere in Polynesia the leaves are fed to pigs. In the Society Islands and Tokelau, the bark is occasionally used in native medicines. In Tonga, hollows in the massive trunks are sometimes used as pig pens. A related inland species, *Pisonia umbellifera*, is common and widespread in Polynesia.

RHIZOPHORA MANGLE
Rhizophoraceae (Mangrove family)

English names: red mangrove, American mangrove
Vernacular names: *tongo* – Tonga; *togo* – Samoa; *dogo* – Fiji

Small tree up to 6 m or more in height, with large, lanceolate stipules on the stems, and numerous stilt roots growing from the branches. Leaves simple, opposite, blade oblanceolate to elliptic, mostly 6–12 cm long, tip slightly notched, margins slightly revolute, surfaces glossy, glabrous, the lower surface black-dotted, petiole 1–2.5 cm long. Flowers in short, axillary, bract-bearing cymes often split into 3 branches at the lowest nodes. Calyx 7–12 mm long, yellow, split into 4 lanceolate lobes. Corolla of 4 brown, deciduous, lanceolate petals 6–9 mm long, hairy on inner surface. Stamens 8. Ovary semi-inferior. Fruit ovoid, 2–3 cm long with the persistent calyx on top, 1-seeded, germinating while still on the plant to produce a long, narrow root. SYNONYM: *Rhizophora samoensis.*

Rhizophora mangle is widely distributed in the New World tropics, but in the Pacific is limited to New Caledonia, Fiji, Tonga, and Samoa, and is introduced and naturalized in Hawai'i. It grows on reef flats, along bays and estuaries, and along the edges of mangrove forest dominated by *Bruguiera gymnorrhiza*, often forming thickets.

Some botanists identify the South Pacific species as *Rhizophora samoensis*, but the differences cited are minor. The wood is often used for firewood and sometimes in light construction. In Fiji and Tonga, a brown dye is obtained from the bark. Its roots were reportedly used for bows in Fiji.

RHIZOPHORA STYLOSA
Rhizophoraceae (Mangrove family)

English name: mangrove
Vernacular names: *tongo* – Tonga; *dogo* – Fiji; *mangle hembra* – Guam

Small to medium-sized tree up to 10 m or more in height, with numerous stilt roots growing from the branches, and large, lanceolate stipules. Leaves simple, opposite, blade coriaceous, elliptic, 6 – 15 cm long, tip with a brittle mucro 2 – 4 mm long, margins slightly revolute, surfaces glabrous, lower surface black-dotted, petiole mostly 15 – 30 mm long. Flowers several, in short, axillary cymes bearing tiny bracts. Calyx 8 – 15 mm long, split into 4 lanceolate lobes, white. Corolla of 4 white, lanceolate petals, 8 – 12 mm long, densely hairy on the margins, deciduous. Stamens 8. Ovary semi-inferior. Fruit ovoid, mostly 2 – 3 cm long, 1-seeded, germinating while still on the plant to produce a long root. SYNONYM: *Rhizophora mucronata* of some authors.

Rhizophora stylosa is distributed from Formosa eastward to Fiji (where it is very common), Tonga (where it is occasional), and the Society Islands (where it was introduced); it has mistakenly been attributed to Samoa. In Micronesia, it is found in the Marianas and Carolines. It grows along estuaries and bays, and on reef flats, often in association with *Rhizophora mangle*.

The same uses are reported for the two *Rhizophora* species, but it is doubtful if the two are distinguished locally. The mucronate tip of *Rhizophora stylosa* is the best differentiating character. A similar species, *Rhizophora apiculata*, is found in Micronesia (Marianas, Carolines).

SCHLEINITZIA INSULARUM
Fabaceae (Pea family)

Vernacular names: *toroire* – Societies, Cooks (Atiu, Ma'uke); *toromiro* – Cooks; *pepe* (?) – Niue; *feifai* – Tonga

Small tree up to 6 m in height. Leaves alternate, bipinnately compound, with 6–13 pairs of pinnae 12–30 cm long, leaflets 15–32, in opposite pairs, oblong, 4–8 mm long, subsessile, surfaces glabrous. Flowers tiny, numerous in axillary, capitate heads 6–12 mm in diameter. Calyx funnel-shaped, 2–3 mm long, 5-lobed. Corolla of 5 oblanceolate petals slightly longer than calyx. Stamens 10, white to pink. Ovary superior. Fruit a strap-shaped pod 6–12 cm long, flattened or somewhat irregularly twisted at maturity, usually containing 4–8 seeds. SYNONYMS: *Leucaena insularum, Leucaena forsteri.*

Schleinitzia insularum is distributed from New Caledonia to Tahiti. Within its Polynesian range, it is found on most of the high archipelagoes except Samoa, but in Micronesia is reported only from the Marianas. It grows on sandy beaches and in coastal thickets, and sometimes along roadsides near the coast.

No significant uses have been reported for this plant, but in the Cook Islands it is sometimes used for firewood and occasionally in house construction. It is very similar to a small introduced tree, *Leucaena leucocephala*, that is common in disturbed places throughout the tropics.

TERMINALIA CATAPPA
Combretaceae (Tropical-almond family)

English name: tropical almond
Vernacular names: *kamani haole*—Hawai'i; *autara'a*—Societies; *kauariki*—Cooks; *selie*—Niue; *talie*—Samoa; *telie*—Tonga; *tavola*—Fiji; *talisai*—Guam

Large tree up to 25 m in height. Leaves simple, alternate, crowded at branch tips, blade obovate, mostly 15–35 cm long, surfaces mostly glabrous, petiole 4–25 mm long. Flowers 50–90, in axillary spikes 8–18 cm long. Calyx white, campanulate, 1.5–2.5 mm long, divided about halfway into 5 triangular lobes. Corolla absent. Stamens 10. Ovary inferior. Fruit drupe-like, compressed-ovoid, mostly 3–8 cm long, usually with a prominent wing along the edge, fibrous within, 1-seeded, red at maturity.

Terminalia catappa is distributed from tropical Asia to Polynesia. It is found on nearly all of the high archipelagoes of Polynesia and Micronesia, but is a modern introduction to eastern (and perhaps western) Polynesia. It grows in littoral forest and is commonly cultivated in villages. In French Polynesia, it seems to have largely replaced the indigenous *Terminalia glabrata*, which is rare there now.

The highly esteemed wood is used to make large gongs, canoes, houses, and other artifacts. The edible kernel of the fruit is eaten only by children because it is laborious to extract, but in time of famine, anyone may eat it. An infusion of the bark is used in the Cook Islands to bathe fractures, and in Tonga and Samoa for treating mouth infections.

TERMINALIA SAMOENSIS
Combretaceae (Tropical-almond family)

Vernacular names: *autara'a* – Societies; *talie* – Samoa; *telie 'a manu* – Tonga; *talisai ganu* – Guam

Small tree up to 5 m or more in height, with yellow-pubescent foliage and petioles. Leaves simple, alternate, crowded at branch tips, blade obovate, 7–20 cm long, surfaces of mature leaves glabrous, petiole 1–3.5 cm long. Flowers 15–30, in axillary spikes 7–15 cm long. Calyx white, campanulate, 2–4 mm long, split about halfway into 5 triangular lobes. Corolla absent. Stamens 10. Ovary inferior. Fruit ovoid, drupe-like, 1.5–2.5 cm long, fibrous within, 1-seeded, red at maturity. SYNONYMS: *Terminalia microcarpa, Terminalia saffordii.*

Terminalia samoensis is distributed from Indonesia to eastern Polynesia. In Polynesia, it is found in the Society Islands, Tuamotus (Makatea), Samoa, and Tokelau, but is reported from all the major archipelagoes of Micronesia. This species is nearly identical to *Terminalia litoralis* of Melanesia, Fiji, and Tonga, and, according to some authors, Micronesia. Perhaps the two should both be considered as one, which would take the name *Terminalia litoralis.* The tree grows on sandy or rocky coasts, and only rarely very far inland.

The tree does not usually grow to a very large size, but the fine timber is used for making canoes, tool handles, and other artifacts. The fruits are much smaller than those of *Terminalia catappa* and are of little use as food.

THESPESIA POPULNEA
Malvaceae (Mallow family)

English name: milo
Vernacular names: *milo* – Hawai'i, Niue, Samoa, Tonga; *miro* – Societies, Cooks; *'amae* – Tahiti; *mulomulo* – Fiji; *banalu – Guam*

Tree up to 12 m in height. Leaves simple, alternate, blade broadly ovate to cordate, 6–22 cm long, surfaces glabrous, petiole 2–5 cm long. Flowers mostly solitary, axillary, on pedicels 1.5–5 cm long. Calyx broadly cup-shaped, entire or shallowly 5-toothed, 7–12 mm long. Corolla of 5 petals 4–6 cm long, pale yellow with purple at the base. Stamens many, monadelphous. Ovary superior. Fruit a subglobose capsule 2–3 cm in diameter, several-seeded, not splitting but eventually disintegrating to release the seeds.

Thespesia populnea is distributed from tropical Africa to Hawai'i, and is found on all the high archipelagoes of Polynesia and Micronesia, but only rarely on atolls. It is native in the western part of its range, but is probably an ancient introduction to parts of eastern Polynesia. It grows in littoral forest, on the margins of mangrove swamps, and along estuaries, but rarely becomes a dominant species.

The fine-grained wood is much favored throughout Polynesia for making items such as boat parts, bowls, paddles, gongs, household articles, and handicrafts. The tree also has medicinal uses: an infusion of the bark is used to treat mouth infections in Tonga and Samoa; various parts are used for treating ailments such as centipede bites and headaches in Tahiti; and the crushed fruit is used in medicines for treating urinary tract infections in the Cook Islands.

TOURNEFORTIA ARGENTEA
Boraginaceae (Borage family)

English name: tree heliotrope
Vernacular names: *tahinu* – Societies; *tai'inu, tau'unu* – Cooks; *toihune* – Niue; *tausuni* – Samoa; *touhuni* – Tonga; *evu, roronibebe* – Fiji; *hunek* – Guam

Small tree up to 5 m or more in height. Leaves simple, alternate and appearing whorled at branch tips, blade fleshy, oblanceolate to obovate, 10 – 20 cm long, densely silky pubescent on both surfaces. Flowers many, in widely branching, paniculate scorpeoid cymes up to 20 cm long. Calyx deeply divided into 5 lobes 1 – 2 mm long. Corolla broadly campanulate, white, 2.5 – 4 mm long, divided about halfway into 5 elliptic lobes. Stamens 5. Ovary superior. Fruit green, globose, 3 – 6 mm long, ultimately dividing into four nutlets. SYNONYMS: *Argusia argentea, Messerschmidia argentea.*

Tournefortia argentea is distributed from Madagascar to the Tuamotus, and is found on most of the low and high islands of Micronesia and Polynesia, but is a modern introduction to Hawai'i. It grows in littoral forest on rocky and sandy coasts, and is particularly common in sandy open habitats of atolls, often being the tree species closest to the ocean.

Since the tree is small, it is not very good for timber, but the wood is sometimes used for making gongs, canoe bailers, tool handles, and carved handicrafts, and parts of the tree are reported to be used in native medicines in the Society Islands and Tokelau. The leaves were once used in the preparation of a red dye in Tahiti.

XYLOCARPUS GRANATUM
Meliaceae (Mahogany family)

English name: puzzle nut
Vernacular names: *lekileki*—Tonga; *legilegi*—Fiji (Lau); *dabi*—Fiji

Medium-sized tree up to 12 m or more in height. Leaves alternate, pinnately compound with a rachis 3—14 cm long, leaflets opposite, usually 4 (2—6), blades obovate to oblong, 6—15 cm long, base unequally-sided, on a short petiole, surfaces glossy, glabrous. Flowers several, in axillary panicles up to 6 cm long. Calyx about 2 mm long, with 4 rounded lobes spreading at maturity. Corolla of 4 white petals 3—6 mm long, spreading at maturity. Stamens 8, filaments fused into a tube. Ovary superior. Fruit a large, hanging, globose capsule 12—25 cm in diameter, brown at maturity, splitting open by 4 valves to release the 8—20 large, corky, irregularly angled seeds. SYNONYMS: *Xylocarpus obovatus, Carapa obovata.*

Xylocarpus granatum is distributed from India eastward to Polynesia (Fiji, Tonga) and Micronesia (Carolines), and grows on the inner edges of mangrove forests and in littoral forests on rocky shores. It differs from the similar *Xylocarpus moluccensis* in having shorter inflorescences, fewer and differently shaped leaflets, and larger fruits, but the two species are probably not distinguished from each other by local people in places where the species occur together.

The wood is sometimes used for boat-building in Fiji. In Tonga, medicines made from the scraped bark, with or without other ingredients, are taken to treat ailments thought to be caused by internal "breaks" (*fasi*) or injuries (*kafo*). It is also sometimes used medicinally in Fiji to treat several different ailments.

XYLOCARPUS MOLUCCENSIS
Meliaceae (Mahogany family)

English name: puzzle nut
Vernacular names: *le'ile'i* – Samoa; *lekileki* – Tonga; *legilegi* – Fiji (Lau); *dabi* – Fiji; *lalanyok* – Guam

Medium-sized tree up to 10 m or more in height. Leaves alternate, pinnately compound with a rachis 8 – 20 cm long, leaflets opposite, 4 – 8, blades ovate, 7 – 16 cm long, subsessile, base unequally-sided, surfaces glossy, glabrous. Flowers many, in axillary panicles 10 – 16 cm long. Calyx about 1.5 mm long, with 4 rounded, spreading lobes. Corolla of 4 white to pale yellow petals 3 – 5 mm long, spreading at maturity. Stamens 8, filaments fused into at tube. Ovary superior. Fruit a hanging, large, globose capsule 7 – 12 cm in diameter, brown at maturity, splitting by 4 valves to release the 8 – 20 corky, irregularly angled seeds. SYNONYM: *Carapa moluccensis*.

Xylocarpus moluccensis is distributed from Madagascar to Polynesia (Fiji, Tonga, Samoa) and Micronesia (Marianas), and grows on rocky and sandy shores in littoral forest and along the margins of mangrove forest. It differs from *Xylocarpus granatum* in having longer inflorescences, more and differently shaped leaflets, and smaller fruits.

In Tonga, the scraped bark, with or without other ingredients, is brewed into a medicine taken for treating "breaks" (*fasi*) and internal injuries (*kafo*). The bark is also sometimes used medicinally in Fiji and Samoa, and the wood is occasionally used for a variety of purposes.

SHRUBS

BATIS MARITIMA
Bataceae (Saltwort family)

English name: pickleweed
Vernacular name: *'ākulikuli kai* – Hawai'i

Small shrub 1 – 1.5 m in height with stems often 4-angled (3 – 5), glabrous, creeping and rooting at the tips to form large colonies. Leaves simple, opposite, blade succulent, linear, curved, mostly 1 – 4 cm long, sessile, surfaces glabrous. Flowers in green or yellow, axillary or terminal, cone-like spikes 4 – 12 mm long, each flower subtended by an overlapping bract and enclosed within a membrane that splits open at the top into 4 lobes, unisexual, plants dioecious. Male flowers with 4 perianth parts, stamens 4. Female flowers lacking perianth parts, ovary superior, with 2 sessile stigmas. Fruits 4-seeded, all fused together to form an irregular, ovoid to oblong, fleshy multiple fruit 8 – 20 mm long.

Batis maritima is native to tropical and subtropical America, and in the Pacific is restricted to Hawai'i, where it was probably unintentionally introduced before 1860. It is thoroughly naturalized on all the main islands in salt marshes and estuary margins, often forming pure stands.

Pickleweed has no reported uses in Hawai'i, but is a dominant species in the saline habitats it occupies, and is hence an important component of the vegetation. The family is represented by a single genus and species, although some botanists recognize another species named from New Guinea.

BIKKIA TETRANDRA
Rubiaceae (Coffee family)

Vernacular names: *siale tafa*—Tonga; *tiale tofa*—Niue; *gaosali*—Guam

Small shrub up to 3 m in height, but usually low and sprawling, with interpetiolar stipules united around the stem. Leaves simple, opposite, blade obovate to elliptic, mostly 5–15 cm long, surfaces glabrous. Flowers solitary, axillary, on a pedicel mostly 10–15 mm long. Calyx deeply divided into 4 linear-lanceolate lobes 2–5 mm long. Corolla funnelform, showy white, 10–15 cm long, divided at the top into 4 spreading, ovate lobes. Stamens 4, exserted. Ovary inferior. Fruit a narrowly ovoid capsule 2–3 cm long, splitting along 4 longitudinal seams to release the numerous angular seeds.

Bikkia tetrandra is distributed from New Caledonia and Micronesia (Marianas) to Fiji and western Polynesia (Niue, Tonga, Horne Islands). It grows in open, sunny places on coastal limestone cliffs and rocky slopes near sea level, sometimes being locally common.

No uses are reported for this plant, although with its showy flowers it would make an attractive ornamental—if it can be cultivated in other habitats.

CAESALPINIA BONDUC
Fabaceae (Pea family)

English names: gray nickers, wait-a-bit
Vernacular names: *kakalaioa* – Hawai'i; *tataramoa* – Societies, Cooks; *talamoa* – Niue; *talatala'amoa* – Tonga; *'anaoso* – Samoa; *soni* – Fiji; *pakao* – Guam

Sprawling, prickly shrub or high-climbing liana with pinnately divided stipules. Leaves alternate, even-pinnately compound, leaflets 6–10 pairs per pinna, blades ovate to elliptic, 1.5–6.5 cm long. Flowers many, in axillary racemes, unisexual, plants monoecious. Calyx 6–9 mm long, deeply divided into 5 unequal sepals. Corolla of 5 yellow, oblong petals 6–9 mm long. Stamens 10, sterile in female flowers. Ovary superior. Fruit a prickly oblong pod 6–9 cm long, containing 1–2 glossy gray, ovoid to globose seeds 15–20 mm in diameter. SYNONYMS: *Caesalpinia bonducella, Caesalpinia crista* of some authors.

Caesalpinia bonduc is pantropical in distribution and widespread in Polynesia (Hawai'i, Easter Island, Society Islands, Austral Islands, Marquesas, Henderson Island, Samoa, Tonga) and Micronesia (all the major archipelagoes). A similar pantropical species often confused with it, *Caesalpinia major*, differs most obviously in its lack of the pinnate stipules. It is widespread in Polynesia (the Australs, Cooks, Niue, Samoa, Tonga, Hawai'i) and Micronesia (Marianas, Carolines, Marshalls). These two species occur in littoral thickets and forests of high islands, up into montane forest in some places.

The attractive seeds are strung into leis and are used as marbles. In Samoa and Tonga, the prickly stems, attached to a stick, are used to snare fruit bats. The nasty, hooked prickles readily rip clothing and skin, and make the plant a pest, especially where it forms thickets.

CAPPARIS CORDIFOLIA
Capparaceae (Caper family)

Vernacular names: *pua pilo* – Hawai'i (*Capparis sandwichiana*); *papiro* – Cooks; *pamoko* – Niue

Woody, prostrate or low shrub up to 1 m in height. Leaves simple, alternate, blade somewhat fleshy, elliptic to ovate, 2–7 cm long, mostly rounded at both ends, petiole 1–4 cm long. Flowers solitary, axillary on a pedicel 5–8 cm long. Calyx of 4 unequal, overlapping sepals 14–25 mm long. Corolla of 4 unequal, showy white petals 2.5–5 cm long. Stamens numerous, anthers often pink. Ovary superior, on a long stipe. Fruit a club-shaped capsule 3–6 cm long, on a long stalk, seeds numerous, globose, 3–4 mm in diameter. SYNONYMS: *Capparis mariana, Capparis spinosa* var. *mariana.*

Capparis cordifolia is widely distributed from Micronesia to eastern Polynesia. In Micronesia, it occurs in the Carolines (Palau), Marshalls (from a questionable record), and Marianas; in Polynesia, it is found in most of the archipelagoes as far east as the Tuamotus, except the Society Islands (it was recorded once from Tahiti, but this record is questionable). It grows on rocky shores, and sometimes along sandy beaches, but rarely occurs very far inland.

This and a closely related Hawaiian species, *Capparis sandwichiana*, are considered by some botanists to be part of a wider-ranging species, *Capparis spinosa* var. *mariana*. The Hawaiian species differs principally in having 120–180 stamens, while *Capparis cordifolia* has less than 100. The plant often has no local names on the islands where it occurs, and few significant uses have been reported for it. In the Cook Islands, it is considered to be poisonous to goats, and on at least one island (Miti'aro) is reportedly sometimes used as a fish poison.

CHENOPODIUM OAHUENSE
Chenopodiaceae (Goosefoot family)

Vernacular names: *ʻāweoweo, ʻāheahea* – Hawai'i

Prostrate to erect shrub up to 3 m in height. Leaves simple, alternate, blade thick and somewhat fleshy, triangular to rhomboid, mostly 1 – 5 cm long, margins 3 – 7-lobed, surfaces densely mealy, gray, petiole mostly 1 – 2.5 cm long. Flowers in small clusters on large, terminal, much-branched panicles. Calyx 3 – 5-lobed, less than 1 mm long. Corolla absent. Stamens 1 – 5. Ovary superior. Fruit a subglobose utricle about 1 mm long, 1-seeded, enclosed within the persistent calyx. SYNONYM: *Chenopodium sandwicheum.*

Chenopodium oahuense is endemic to Hawai'i, where it grows on the Leeward Islands and all the main islands except Kaho'olawe. It is found in dry, sandy, littoral and coastal habitats, but also inland up to 2500 m in elevation in dryland forest (especially at mid-elevations on the island of Hawai'i).

No significant uses have been reported for this species, but it is a major component of some types of vegetation. Several introduced weedy species of the genus are also found in Hawai'i.

CLERODENDRUM INERME
Verbenaceae (Verbena family)

Vernacular names: *aloalo tai* – Samoa; *tutu hina* – Tonga; *vere, verevere* – Fiji; *lodigao* – Guam

Spreading shrub up to 4 m in height, or sometimes scandent and climbing into trees. Leaves simple, opposite, blade ovate to elliptic, 5–13 cm long, surfaces glabrous, petiole 1–2 cm long. Flowers 1–several, in axillary cymes, on a peduncle 1–5 cm long. Calyx campanulate, 3–6 mm long, margin entire or with 5 tiny lobes. Corolla salverform, white, tube 2.5–3.5 cm long, limb deeply divided into 5 oblong lobes 6–10 mm long. Stamens in 2 pairs, long-exserted. Ovary superior. Fruit obovoid, 1–1.5 cm long, shallowly 4-lobed, splitting at maturity into four 1-seeded nutlets.

Clerodendrum inerme is distributed from tropical Asia to Niue, and is found on most of the high islands of western Polynesia and all the major archipelagoes of Micronesia. It grows on rocky or sandy shores in coastal thickets, along mangrove swamps, on the seaward margin of littoral forest, or climbs as a liana into the canopy of littoral forest.

No significant uses are reported for this plant, but it is sometimes an important part of the littoral vegetation in which it occurs. In Tonga, a plant known as *tuamea,* which has long flexible stems used to make cane chairs, may be this species.

COLUBRINA ASIATICA
Rhamnaceae (Buckthorn family)

Vernacular names: *'anapanapa, kukuku* – Hawai'i; *tutu* – Societies, Cooks; *fihoa* – Tonga, Niue; *fisoa* – Samoa; *vere, vusolevu* – Fiji; *gasoso* – Guam

Shrub or sprawling liana with stems often somewhat zigzag. Leaves simple, alternate, blade ovate, 3 – 12 cm long, 3-veined from base, surfaces glossy grccn, mostly glabrous, petiole 7 – 18 mm long. Flowers tiny, in short, axillary cymes 5 – 10 mm long. Calyx of 5 triangular lobes 1 – 2 mm long. Corolla of 5 yellow petals 1 – 2 mm long. Stamens 5. Ovary inferior. Fruit a globose capsule 6 – 10 mm in diameter, brown and papery at maturity, 3-seeded.

Colubrina asiatica is distributed from tropical East Africa to Hawai'i, and is found on all the major high islands of Polynesia and in the Carolines and Marianas of Micronesia. It grows in littoral thickets on rocky and sandy shores of high islands, and sometimes climbs as a liana into the canopy of littoral forests.

The roots, bark, and leaves contain saponins used in ancient times for soap throughout Polynesia. In the Cook Islands, the flexible stems are used to weave fish traps and as lashing on goatskin drums. In the Society Islands, the leaves are occasionally used in native medicines.

CORCHORUS TORRESIANUS
Tiliaceae (Linden family)

Vernacular names: none

Shrub up to 1 m in height, with velvety pubescent stems. Leaves simple, alternate, blade obovate, 2–4 cm long, toothed on the rounded or truncate tip, gray and velvety pubescent on both surfaces. Flowers 1–3, in short, axillary cymes. Calyx 6–9 mm long, deeply divided into 5 lanceolate lobes. Corolla of 5 yellow petals 6–9 mm long. Stamens many, showy yellow. Ovary superior. Fruit an ellipsoid capsule 12–20 mm long, splitting open at the top into 5 valves, covered with short, thick, stiff bristles.

Corchorus torresianus is native from New Caledonia and Micronesia eastward to the Tuamotus. In Polynesia, it is reported only from the Tuamotus (Ana'a), Cooks, Tonga (Tongatapu), and Fiji, and in Micronesia is found in the Marianas, but not on Guam. It grows on rocky limestone shores or sandy beaches on atolls and high islands, but nowhere in its range is it reported to be common.

Because the plant is so infrequent, no names or uses have been reported for it within its Polynesian range.

DENDROLOBIUM UMBELLATUM
Fabaceae (Pea family)

Vernacular names: *lala* – Samoa; *lala'uta* – Tonga; *tokaibebe* – Fiji; *palaga hilitai* – Guam

Shrub mostly 1 – 4 m in height. Leaves alternate, trifoliate, leaflet blades mostly oblong to elliptic, 5 – 12 cm long, surfaces glabrous. Flowers several, in short axillary clusters bearing small deciduous bracts. Calyx campanulate, 4 – 6 mm long, divided about halfway into 4 ovate lobes, pubescent, subtended by a pair of deciduous bracteoles. Corolla papilionaceous, white to pale yellow, 8 – 12 mm long. Stamens 10, diadelphous. Ovary superior. Fruit a pod constricted about halfway into 1 – 5 oblong, 1-seeded segments 7 – 11 mm long, brown and separating at maturity. SYNONYM: *Desmodium umbellatum.*

Dendrolobium umbellatum, until recently known as *Desmodium umbellatum*, is distributed from East Africa to western Polynesia (Fiji, Tonga, Samoa, Horne Islands, Wallis, Niue) and Micronesia (Marianas, Carolines). It is common in beach thickets of high islands (rarely on atolls) and along the margins of mangrove swamps and estuaries, and sometimes grows in open places inland up to 200 m in elevation.

No significant uses have been reported for this plant, except for its inclusion in native medicines (e.g., in Tonga). This reported medicinal use may be in error, however, because the same common name (*lala*) is applied to *Vitex trifolia*, a commonly used medicinal plant in western Polynesia.

EUGENIA REINWARDTIANA
Myrtaceae (Myrtle family)

Vernacular names: *nīoi* – Hawai'i, Societies, Cooks; *liki* – Niue; *unuoi* – Samoa?, Tonga; *a'abang* – Guam

Shrub or small tree up to 7 m in height. Leaves simple, opposite, blade coriaceous, mostly elliptic to ovate, 3–8 cm long, margins often revolute, glossy green above, dull green below, petiole 2–4 mm long. Flowers 1 or 2, axillary, on pedicels 5–25 mm long, with a pair of small terminal bracts. Calyx of 4 rounded, unequal sepals 2–5 mm long. Corolla of 4 white, ovate petals 6–8 mm long, deciduous. Stamens many, white. Ovary inferior. Fruit a fleshy, subglobose berry 1–2 cm long, red at maturity, containing 1–4 seeds. SYNONYMS: *Eugenia rariflora, Jossinia reinwardtiana.*

Eugenia reinwardtiana is distributed from Borneo and Micronesia to Hawai'i, and occurs on all the major high archipelagoes of Polynesia (except perhaps the Marquesas) and the Marianas and Carolines of Micronesia. It commonly grows as a low shrub on coastal rocks and makatea coasts, but sometimes is taller in coastal forest. In Hawai'i, where it is uncommon, it grows inland in dry forest.

The edible fruit is reportedly eaten over much of its range, mostly by children, but by everyone in times of famine. The leaves are used in native medicines in Tonga.

FICUS SCABRA
Moraceae (Mulberry family)

Vernacular names: *masi* – Niue, Tonga; *mati* – Samoa; *nunu, masi* – Fiji

Shrub or small tree up to 10 m or more in height, with milky latex. Leaves simple, alternate, blade ovate to elliptic, 5 – 20 cm long, base rounded to nearly cordate or oblique, margins wavy to entire, surfaces rough. Flowers minute, enclosed within a fruit-like structure called a syconium. Calyx of 2 – 8 tiny sepals. Corolla absent. Stamens 1 or 2. Ovary superior. Fruit a globose, berry-like multiple fruit (syconium) 8 – 15 mm in diameter, many-seeded, red to yellow at maturity, on a stalk mostly 5 – 10 mm long, solitary or in clusters on branchs or the trunk.

Ficus scabra is distributed from New Caledonia to Niue, and is found on most of the high islands in this region, but does not extend into Micronesia. It grows as a shrub on windswept coastal slopes and in coastal thickets, and often as an understory tree in coastal to lowland forest up to 300 m in elevation.

Few significant uses are reported for the plant, but in Fiji the young cooked leaves are reportedly edible. The rough-surfaced leaves may also be used as sandpaper. A related species, *Ficus tinctoria*, which ranges from India to Tahiti, differs from *Ficus scabra* in having smooth, relatively broader leaves rather than scabrous ones, and is sometimes cultivated on atolls for its edible fruits.

GOSSYPIUM HIRSUTUM
Malvaceae (Mallow family)

English name: Polynesian cotton
Vernacular names: *vavai* – Societies; *vavae* – Samoa; *vauvau* – Fiji

Shrub up to 4 m in height, with foliage and other parts covered with tiny black glands. Leaves simple, alternate, blade usually palmately 3-lobed, 3–12 cm long and nearly as wide, surfaces mostly glabrous, black-dotted on both sides, petiole 2–6 cm long. Flowers axillary, solitary, on a pedicel 2–4 cm long. Calyx shallowly cup-shaped, slightly lobed, 6–10 mm long. Corolla of 5 petals 3–5 cm long, yellow to cream-colored and red at the base. Stamens numerous, monadelphous. Ovary superior. Fruit a 3–5-valved, globose to ovoid, beaked capsule 1.5–2.5 cm long, with 3 deeply toothed, persistent bracts below it, seeds 5–11 per cell, densely covered with lint. SYNONYM: *Gossypium religiosum*.

Gossypium hirsutum is native to tropical America, but one variety, var. *taitense*, is indigenous to the Society Islands, Tuamotus, Marquesas, Fiji, Samoa, and Micronesia. It grows on windswept coastal ridges and in scrubby vegetation near the shore, but occasionally inland up to 400 m elevation (Marquesas).

This species of cotton was of little reported use in ancient Polynesia, but introduced varieties and species were once an important commercial crop there in the nineteenth century. A related species endemic to Hawai'i, *Gossypium tomentosum*, differs most obviously in having bright yellow flowers.

LYCIUM SANDWICENSE
Solanaceae (Nightshade family)

Vernacular name: *'ōhelo kai* – Hawai'i

Low, glabrous shrub up to 1 m in height. Leaves simple, arranged in alternate fascicles on the stem, blade succulent, spathulate, 5–30 mm long, subsessile, surfaces glabrous. Flowers solitary from the fascicles, on a pedicel 5–15 mm long. Calyx 2–4 mm long, divided about halfway into 4 triangular lobes. Corolla pink to white, 5–7 mm long, divided into 4 rounded lobes. Stamens 4. Ovary superior, style 2–lobed. Fruit a red, subglobose berry 6–10 mm long. SYNONYM: *Lycium carolinense* var. *sandwicense.*

Lycium sandwicense is endemic to Polynesia, ranging from Tonga to Hawai'i and Easter Island, and within this area is found in scattered localities (Australs, Mangareva, Pitcairn, Henderson). It grows in sunny places on rocky shores of high islands or makateas, but is apparently absent from atolls.

Few uses are reported for this plant, because over most of its range it is uncommon or rare. On Rapa (Australs), Hawai'i, and probably other islands, the bright red berries are sometimes eaten.

MYOPORUM SANDWICENSE
Myoporaceae (Myoporum family)

English name: false sandalwood
Vernacular names: *naio* – Hawai'i; *ngaio* – Cooks

Shrub to small tree 1–10 m or more in height. Leaves simple, alternate, blade somewhat fleshy, elliptic to narrowly lanceolate, mostly 4–15 cm long, margins entire or serrate (at least when young), surfaces glabrous, petiole 1–25 mm long. Flowers solitary, or 1–several in short axillary cymes, on a pedicel 5–17 mm long. Calyx 2–5 mm long, deeply divided into 5 ovate to lanceolate lobes. Corolla campanulate to funnelform, white, 5–10 mm long, 5-lobed. Stamens 5. Ovary superior. Fruit a subglobose to ovoid drupe mostly 4–8 mm long, several-seeded, flesh-colored at maturity. SYNONYM: *Myoporum wilderi*.

Myoporum sandwicense is found in Hawai'i, where it is reported from all the main islands, and in the Cook Islands (Mangaia, Miti'aro, Rarotonga). It grows near the shore on rocky or sandy coasts, but in Hawai'i, where several varieties have been recognized, it also occurs as a small tree in dryland forest at high elevations (to over 2000 m elevation). Two other very similar species are recorded from Polynesia, *Myoporum rapense* and *Myoporum stokesii* from the Austral Islands, but these may eventually be considered to be the same species as *Myoporum sandwicense*. Another similar species, *Myoporum boninense*, occurs in the Marianas.

The somewhat fragrant wood of this tree once served as a poor substitute for sandalwood in Hawai'i and was reportedly once used for house frames there. On Mangaia, the flowers are used for scenting coconut oil, and in ancient times, it was reportedly employed in mummifying corpses.

PEMPHIS ACIDULA
Lythraceae (Loosestrife family)

English name: pemphis
Vernacular names: *'a'ie* – Societies; *ngangie* – Cooks; *ngingie* – Niue, Tonga; *gigia* – Fiji; *nigas, nietkot* – Guam

Shrub or small tree up to 5 m or more in height. Leaves simple, opposite, blade oblanceolate, 1 – 2.5 cm long, subsessile, appressed-pubescent on both surfaces. Flowers usually solitary in the leaf axils, pedicel 5 – 15 mm long. Calyx campanulate, 3 – 7 mm long, shallowly 6-lobed. Corolla of 6 white, wrinkled, elliptic to ovate petals 4 – 8 mm long. Stamens 12. Ovary superior. Fruit a reddish, obovoid to ellipsoid capsule 5 – 10 mm long, enclosed within the calyx tube, opening by means of a cap to release the numerous seeds.

Pemphis acidula is distributed from tropical East Africa to eastern Polynesia, and is found on all the major high and low archipelagoes of Micronesia and Polynesia except Hawai'i and the Marquesas. It grows on coral sand or rock on atolls and makatea islands, but only occasionally on rocky shores of high volcanic islands (e.g., it is rare in Samoa).

The tree does not grow to a large size, but the hard wood is used for house parts, tool handles, fish hooks, and boat parts in Polynesia and Micronesia. In Tokelau, the roots and bark are used to make a red dye. The tree is usually not distinguished from *Suriana maritima*, which often goes by the same local names.

PHYLLANTHUS SOCIETATIS
Euphorbiaceae (Spurge family)

Vernacular names: none

Small subshrub up to 1 m in height. Leaves simple but appearing pinnately compound, alternate, arranged in a single plane, blade obovate, 1–2 cm long, surfaces glabrous, petiole about 1 mm long. Flowers unisexual, plants dioecious. Male flowers in dense axillary clusters on the lower side of the branches, female flowers solitary in the leaf axils. Calyx of 4–6 rounded, green sepals about 1 mm long. Corolla absent. Stamens 2–5. Ovary superior, 3-celled. Fruit a green, subglobose capsule 2–3 mm in diameter, 6-seeded.

Phyllanthus societatis is endemic to eastern Polynesia, and is reported only from the Tuamotus (Makatea), Society Islands (Mehetia), and the Cook Islands (Atiu, Ma'uke, Miti'aro). It grows in open sandy places on the shore, or sometimes inland in forest clearings and along paths. Several other endemic species of this genus occur in Polynesia and Micronesia, but these are mostly inland species except for *Phyllanthus marianus* of the Marianas, which is nearly indistinguishable from *Phyllanthus societatis*. Several weedy species of this genus are also present in the islands.

No uses or local names are reported for this plant over its native range.

PREMNA SERRATIFOLIA
Verbenaceae (Verbena family)

Vernacular names: *'avaro*—Societies; *aloalo*—Niue, Samoa; *volovalo*—Tonga; *yaro*—Fiji; *ahgao*—Guam

Shrub or small tree up to 7 m in height. Leaves simple, opposite, blade ovate to elliptic, 5–18 cm long, base acute to cordate, surfaces glabrous, petiole 5–30 mm long. Flowers many, in dense, repeatedly branching, terminal panicles. Calyx cup-shaped, 1–2 mm long, shallowly 4-lobed. Corolla slightly bilabiate, white, 2–3 mm long. Stamens 4. Ovary superior. Fruit a fleshy, black, globose to ovoid drupe 3–7 mm long, 4-seeded, with a persistent, enlarged, saucer-shaped calyx at the base. SYNONYMS: *Premna integrifolia, Premna taitensis* of some authors, *Premna obtusifolia, Premna gaudichaudii*

Premna serratifolia is distributed from tropical Asia to the Marquesas. It is found on all the high archipelagoes of Micronesia and Polynesia except Hawai'i, growing mostly on rocky shores of high islands and on the larger atolls of the Tuamotus, and sometimes inland in open forest and on ridges. Some botanists distinguish several species, but the differences between these are obscure.

The only reported use for *Premna serratifolia* is as an ingredient in native medicines. Its leaves are somewhat bad-smelling, and are used in treating "ghost sickness" (supernaturally induced ailments) in the Society Islands, and for treating infections and sores in Tonga and Samoa.

SCAEVOLA CORIACEA
Goodeniaceae (Goodenia family)

English name: dwarf naupaka
Vernacular name: *naupaka* – Hawai'i

Prostrate, perennial plant woody at the base. Leaves simple, opposite, blade succulent, obovate to spathulate, 2–5 cm long, margins entire, surfaces glabrous, petiole 7–15 mm long. Flowers 1–3, in axillary cymes 1–2 cm long, subtended by a pair of lanceolate bracteoles 1–2 mm long. Calyx entire, about 1 mm long. Corolla funnelform, yellowish green, hairy inside, 15–20 mm long, divided about halfway into 5 spreading linear lobes, and split to the base on the upper side. Stamens 5. Ovary inferior, style 2-lobed. Fruit an ovoid drupe 5–10 mm long, purplish black, 1-seeded.

Scaevola coriacea is endemic to Hawai'i, where it is reported from all the main islands except Kaho'olawe. However, it may now be extinct throughout its range except for an offshore islet of Moloka'i, and on Maui, where it grows in a few scattered localities, mostly on stabilized sand dunes on the coast, and is on the U.S. Federal list of endangered species.

No uses are reported for this species, and its Hawaiian name is a general one for members of the genus. In addition to *Scaevola coriacea*, Hawai'i has seven other endemic species, all of them occurring in inland habitats, as well as the widespread *Scaevola taccada*, which is common in littoral habitats.

SCAEVOLA TACCADA
Goodeniaceae (Goodenia family)

English name: scaevola
Vernacular names: *naupaka kahakai* – Hawai'i; *naupata* – Societies; *nga'u* – Cooks; *ngahupā* – *Niue; ngahu* – Tonga; *to'ito'i* – Samoa; *veveda* – Fiji; *nanasu* – Guam

Shrub up to 3 m in height, with leaf axils bearing a tuft of white hairs. Leaves simple, alternate, crowded at stem tips, blade somewhat fleshy, spathulate or oblanceolate to suborbicular, 5–20 cm long, margins revolute, surfaces glossy, glabrous. Flowers several, in short axillary cymes 1–4 cm long. Calyx 1–6 mm long, deeply divided into 5 narrowly lanceolate lobes. Corolla white to pale yellow, 1.5–2.3 cm long, split along one side and 5-lobed. Stamens 5. Ovary inferior. Fruit a fleshy white, subglobose drupe 8–15 mm long. SYNONYMS: *Scaevola frutescens, Scaevola koenigii, Scaevola sericea.*

Scaevola taccada is distributed from India to Hawai'i, and is found on all the major archipelagoes of Polynesia and Micronesia. It is one of the most common littoral shrubs, often forming dense thickets on rocky and sandy coasts throughout the area. A prostrate variety with yellowish flowers found in eastern Polynesia has been called var. *tuamotuense.*

The fruits of scaevola are eaten by pigeons, as well as by sea birds, especially on atolls where few other suitable fruits are available. In several Polynesian archipelagoes, the leaves are used in native medicines. The pith from the stems is strung into leis in Tokelau, and is made into dancing skirts in the Cook Islands.

SESBANIA TOMENTOSA
Fabaceae (Pea family)

Vernacular name: *'ōhai*—Hawai'i

Prostrate shrub to small tree up to 6 m in height. Leaves alternate, even-pinnately compound on a rachis 6–30 cm long, leaflets 18–38, opposite, blades oblong to narrowly elliptic, 15–35 mm long, surfaces densely silvery tomentose. Flowers 2–9, in axillary racemes 1–7 cm long. Calyx cup-shaped, 7–12 mm long, shallowly divided into 5 triangular lobes, pedicel 1–4 cm long. Corolla papilionaceous, yellow to red, 2.3–4.5 cm long. Stamens 10, diadelphous. Ovary superior. Fruit a linear, somewhat flattened pod 7–23 cm long, slightly constricted between the 6–25 seeds that are 4–6 mm long. SYNONYMS: *Sesbania molokaiensis, Sesbania hawaiiensis, Sesbania arborea, Sesbania hobdyi.*

Sesbania tomentosa is endemic to Hawai'i, where it has been reported from all the main islands of the archipelago. However, it is now restricted to scattered relictual populations on sandy beaches and sometimes inland in open vegetation up to 800 m in elevation.

No significant uses are reported for this plant in Hawai'i. Some botanists recognize several varieties or species. *Sesbania tomentosa* is related to *Sesbania atollensis* of eastern Polynesia and *Sesbania coccinea* of New Caledonia (Isle of Pines), Fiji, and Tonga (possibly extinct there now).

SIDA FALLAX
Malvaceae (Mallow family)

Vernacular name: *'ilima* – Hawai'i

Prostrate to erect shrub up to 1.5 m or more in height. Leaves simple, alternate, blade ovate to suborbicular, mostly 3 – 9 cm long, petiole about half as long, margins toothed, lower surface velvety pubescent. Flowers solitary, axillary, on a stalk up to 7 cm long in fruit. Calyx campanulate, 6 – 12 mm long, divided about halfway into 5 triangular lobes. Corolla rotate, pale orange, 8 – 15 mm long, divided into 5 broadly ovate petals unequally lobed at the tip. Stamens many, monadelphous. Ovary superior. Fruit a schizocarp 2.5 – 4 mm long, splitting at maturity into 6 – 9 beaked segments. SYNONYMS: *Sida cordifolia* of some authors, *Sida meyeniana*.

Sida fallax is distributed from China to eastern Polynesia. In Micronesia, it is found in the Carolines, Gilberts, and Marshalls, and in Polynesia, it occurs in Hawai'i, the Society Islands (Teti'aroa), and the Marquesas. It grows on sandy beaches and rocky shores, and inland in open forest up to an elevation of over 750 m (on Hawai'i). It has often been confused with *Sida cordifolia*, which is a naturalized weed found in Tonga, the Tuamotus (Makatea), Mangareva, and Hawai'i.

In Hawai'i, where it is common, the flowers are highly esteemed for making leis that are sold commercially. In former times, the chewed flowers were given to infants as a mild laxative.

SOPHORA TOMENTOSA
Fabaceae (Pea family)

Vernacular names: *pofatu'ao'ao* – Societies; *po'utukava* – Cooks

Shrub up to 6 m in height. Leaves alternate, odd-pinnately compound, rachis up to 30 cm or more long, leaflets opposite, mostly 13–19, blades oblong to nearly round, 2–5 cm long, margins entire, upper and lower surfaces pubescent, gray-green. Flowers many, in axillary or terminal racemes 10–30 cm long. Calyx cup-shaped, 5–9 mm long, shallowly 4-toothed, pubescent. Corolla papilionaceous, yellow, 13–21 mm long. Stamens 10, diadelphous. Ovary superior. Fruit a pod 7–15 cm long, greatly constricted between the 2–8 subglobose seeds.

Sophora tomentosa is pantropical in distribution, and is found in most of the high archipelagoes of Polynesia as far east as the Marquesas (possibly extinct there now), and on all the major archipelagoes of Micronesia. It grows mostly on sandy beaches, rarely far from the shore. Two related species occur in Polynesia – *Sophora tetraptera* of Chile, New Zealand, and Rapa, and *Sophora chrysophylla* of Hawai'i – but both of these are inland forest trees.

Although conspicuous, in much of Polynesia the plant lacks vernacular names. In the Cook Islands, however, its seeds are sometimes strung into leis, the leaves are employed in native medicines, and the stems are burned for firewood.

SURIANA MARITIMA
Surianaceae (Suriana family)

Vernacular names: *'o'uru* – Societies; *kuru* – Cooks (Penrhyn); *ngangie* – Cooks; *ngingie* – Tonga; *gigia* – Fiji; *nietkot* – Guam

Shrub up to 4 m in height. Leaves simple, alternate, crowded at branch tips, blade narrowly lanceolate, 1.5 – 3 cm long, subsessile, surfaces appressed-pubescent. Flowers several, in axillary cymes 1 – 3 cm long. Calyx 6 – 9 mm long, deeply divided into 5 lanceolate lobes. Corolla rotate, of 5 obovate to suborbicular, yellow petals about as long as calyx. Stamens 10, in 2 series. Ovary superior. Fruit composed of 3 – 5 ovoid drupes 2 – 4 mm in diameter, pubescent, dry, splitting apart at maturity.

Suriana maritima is pantropical in distribution, and in Polynesia is found on most of the high and low archipelagoes eastward as far as the Tuamotus, and on all the major archipelagoes of Micronesia. It grows on sandy or coral rubble beaches, or on makatea coasts, but rarely very far from the coast. It is particularly common in the Tuamotus, and is rare in western Polynesia.

This shrub is superficially similar to *Pemphis acidula*, and is often mistaken for it. In the Tuamotus, the hard wood is used for fish hooks, the scraped inner bark is used to caulk canoes, and the leaves are sometimes employed in native medicines.

TIMONIUS POLYGAMUS
Rubiaceae (Coffee family)

Vernacular names: *kōpara* – Cooks (Miti'aro, Ma'uke); *kaveutu* – Niue

Prostrate or erect shrub up to 5 m in height, with interpetiolar stipules. Leaves simple, opposite, blade coriaceous, obovate to suborbicular, 3–10 cm long, surfaces glabrous, petiole up to 1 cm long. Flowers unisexual, plants dioecious. Male flowers in axillary cymes up to 5 cm long, female flowers usually solitary, axillary. Calyx campanulate, 2–5 mm long, shallowly 4–5-toothed. Corolla funnelform, white, 5–9 mm (female) or 12–18 mm (male) long, 4–6-lobed. Stamens as many as corolla lobes. Ovary inferior. Fruit a subglobose berry 1–1.5 cm long, purple to black, several-seeded, with a persistent calyx. SYNONYM: *Timonius forsteri.*

Timonius polygamus is distributed from the Solomons (Rennell Island) eastward to the Tuamotus (Henderson Island), and is found in most of the archipelagoes in this area, except Samoa. It grows on sandy or makatea shores, and sometimes inland in undisturbed habitats on makatea islands. It is superficially similar to some species of *Xylosma*, with which it is sometimes confused.

There are few reported uses for this plant, other than its fruit being food for pigeons.

VITEX ROTUNDIFOLIA
Verbenaceae (Verbena family)

English name: beach vitex
Vernacular name: *pōhinahina* – Hawai'i

Low, widely branching shrub with decumbent stems often rooting at the nodes and forming large mats. Leaves simple (or rarely palmately compound), opposite, blade obovate to suborbicular, 2–6.5 cm long, upper leaf surface green, lower pale green and densely tomentose, subsessile or with a petiole up to 1 cm long. Flowers several, usually in terminal cymes 3–7 cm long. Calyx cup-shaped, 3–5 mm long, gray, shallowly divided into 5 triangular lobes. Corolla narrowly funnelform, bilabiate, bluish purple, 9–13 mm long. Stamens in 2 pairs, exserted. Ovary superior. Fruit a subglobose capsule 5–7 mm in diameter, black at maturity, 4-seeded. SYNONYMS: *Vitex ovata, Vitex trifolia* var. *simplicifolia, Vitex trifolia* of some authors.

Vitex rotundifolia is widespread from Mauritius in the Indian Ocean to Hawai'i, but is not found in Micronesia or elsewhere in Polynesia. In Hawai'i, it is reported from all the main islands (except Kaho'olawe) growing on sand dunes, sandy beaches, and sometimes on rocky shores, but it is uncommon and restricted to scattered localities.

The species was occasionally employed in native medicines, but was otherwise rarely used in Hawai'i, where six local names are recorded for it. It is related to *Vitex trifolia*, but is quite different in leaf shape and growth form. Its showy, bluish purple flowers and low habit make *Vitex rotundifolia* an attractive ornamental plant.

VITEX TRIFOLIA
Verbenaceae (Verbena family)

Vernacular names: *rara* – Cooks; *lala sea* – Niue; *lala tahi* – Tonga; *namulega* – Samoa; *drala sala* – Fiji; *lagunde?* – Guam

Shrub or small tree up to 5 m in height, with densely tomentose stems. Leaves opposite, simple or palmately lobed into 3–5 ovate to lanceolate leaflets 1–12 cm long, upper surface green, lower surface tomentose and gray-green. Flowers in terminal, narrow panicles up to 18 cm long. Calyx campanulate, 1–2 mm long, 4–6-lobed. Corolla bilabiate, violet, 4–7 mm long. Stamens in 2 pairs. Ovary superior. Fruit a fleshy, subglobose drupe 5–7 mm in diameter, black at maturity, 4-seeded. SYNONYM: *Vitex negundo* var. *bicolor.*

Vitex trifolia is distributed from tropical East Africa eastward as far as the Marquesas (possibly extinct there now), and is found on most of the high islands in this area and all of the major archipelagoes of Micronesia. It grows in coastal thickets on high islands, but is rare on atolls. It is also sometimes planted around houses, and is grown as an ornamental in Hawai'i. The variety in Polynesia is called var. *bicolor.*

The only uses reported for this plant, other than ornamental, are medicinal. The boiled leaves are used in the Cook Islands to treat postpartum ailments of women. In Samoa, the leaves are used to treat fever, and in Tonga to treat "ghost sickness" (supernaturally induced ailments).

XIMENIA AMERICANA
Olacaceae (Olax family)

Vernacular names: *rama* – northern Cooks; *moli tai* – Samoa; *moli tahi* – Tonga (Niuatoputapu); *vītahi* – Tonga; *somisomi, tomitomi* – Fiji; *pi'ut* – Guam

Erect shrub up to 3 m in height, with sharp spines in the axils. Leaves simple, alternate or in fascicles, blade elliptic, 2.5 – 10 cm long, margins wavy to entire, surfaces glabrous. Flowers several, in short axillary clusters. Calyx cup-shaped, about 1 mm long, deeply 4-lobed. Corolla of 4 white petals 7 – 9 mm long, densely hairy inside. Stamens 8. Ovary superior. Fruit an ovoid drupe 18 – 28 mm long, with a thin yellow pericarp. SYNONYM: *Ximenia elliptica.*

Ximenia americana is pantropical in distribution. In Polynesia, it is found in the Society Islands (possibly extinct there now), Tuamotus, Cooks (Penrhyn), Samoa, Tonga, and Tuvalu; in Micronesia, it occurs in the Marianas, Carolines and Marshalls. It grows on rocky or sandy shores, rarely far from the beach, but is now generally uncommon throughout Polynesia.

The thin, edible pericarp is eaten, mostly by children. The hard wood is used to make the large *Ruvettus* (oil fish) hooks on Penhryn (Cook Islands).

XYLOSMA ORBICULATUM
Flacourtiaceae (Flacourtia family)

Vernacular names: *liki*—Niue; *fululupe*—Tonga

Low, spreading shrub up to 1 m in height. Leaves simple, alternate, blade obovate to elliptic, 2–4.5 cm long, margins slightly revolute, surfaces glabrous, petiole 2–5 mm long. Flowers several, in short axillary racemes 3–10 mm long, unisexual, plants dioecious. Calyx deeply divided into 4 green, broadly ovate sepals 1.5–2.5 mm long. Corolla absent. Stamens of male flowers numerous, red, exserted. Ovary of female flowers superior. Fruit a dry, subglobose berry 7–10 mm long, few-seeded, purple at maturity.

Xylosma orbiculatum is found in Fiji (where it was collected only once), Tonga (where it is uncommon), and Niue (where it is occasional to common), and grows in low vegetation on rocky shores.

No significant uses are reported for this species, except in Tonga, where it is occasionally employed in making dance skirts. Two related species occur in the area, *Xylosma suaveolens* of eastern Polynesia, and *Xylosma simulans* of Fiji and Tonga, but these are trees of inland forests. A third species, *Xylosma samoense*, is endemic to Samoa, where it is found in high-elevation rain forest.

HERBS

ACHYRANTHES SPLENDENS
Amaranthaceae (Amaranth family)

Vernacular names: none

Erect woody herb or subshrub up to 2 m in height. Leaves simple, opposite, blade obovate to nearly round, mostly 2–6 cm long, the surfaces and stems densely white-pubescent, petiole 5–15 mm long. Flowers many, in narrow terminal spikes 3–25 cm long, rachis densely woolly. Calyx of 3 lanceolate sepals 5–8 mm long, with 3 shorter, spine-tipped bracts below. Corolla absent. Stamens 2–5. Ovary superior. Fruit a narrowly ovoid utricle 1.5–3 mm long, 1-seeded, falling enclosed within the persistent bracts and calyx.

Achyranthes splendens is endemic to Hawai'i, where it is known only from scattered localities on O'ahu, Maui, Moloka'i, and Lana'i. It grows in littoral and coastal areas on rocky slopes and coralline plains, and occasionally inland up to 500 m in elevation.

No uses are reported for the plant, nor are any Hawaiian names recorded. Two other related species are endemic to Hawai'i–*Achyranthes mutica* from Kaua'i and Hawai'i, and *Achyranthes atollensis* from the Leeward Hawaiian Islands–but both of these are thought to be extinct. One of the two varieties of *Achyranthes splendens* is on the U.S. Federal list of endangered plants as well. All of these are similar to *Achyranthes velutina* of Polynesia.

ACYHRANTHES VELUTINA
Amaranthaceae (Amaranth family)

Vernacular names: *'aerofai* – Societies; *tāmakomako* – Cooks (Pukapuka)

Erect woody herb or subshrub up to 2 m in height. Leaves simple, opposite, blade elliptic, mostly 5 – 15 cm long, base acute to attenuate, surfaces and stems woolly-pubescent, petiole up to 25 mm long. Flowers many, in narrow, terminal spikes up to 30 cm long. Calyx of 5 hard, green, lanceolate sepals 4 – 7 mm long with a pair of bracteoles below. Corolla absent. Stamens 5, pink. Ovary superior. Fruit an ovoid utricle falling enclosed within the spikelet-like calyx up to 7 mm long, 1-seeded.

Achyranthes velutina is endemic to Polynesia, where it is found in Tokelau, American Samoa (Swains Island), the Society Islands, Cook Islands, Tuamotus, and Australs. Unlike the related, somewhat weedy *Achyranthes aspera*, this species is found almost exclusively in sandy littoral habitats or in clearings on undisturbed makatea.

Achyranthes is of little use to the Polynesians, except, perhaps, occasionally being employed in native medicines. Its sharp fruits, which stick to clothing, make it somewhat of a nuisance where it does occur.

ATRIPLEX SEMIBACCATA
Chenopodiaceae (Goosefoot family)

English name: Australian saltbush
Vernacular names: none

Perennial herb with prostrate stems up to 1.5 m long, forming spreading mats from a long taproot. Leaves simple, alternate, blade elliptic to ovate or spathulate, mostly 5–20 mm long, margins entire to several-toothed, surfaces mealy, petiole short. Flowers many, tiny, in axillary clusters, unisexual, plants monoecious. Calyx tiny, with 3–5 lobes. Corolla absent. Stamens 3–5 in male flowers. Ovary of female flower superior, subtended by 2 bracts. Fruit a 1-seeded utricle surrounded by the 2 red, rhomboid, fleshy bracts 4–6 mm long.

Atriplex semibaccata is native to Australia, but was apparently introduced to Hawai'i as a trial forage crop before 1900 and has become naturalized on all the main islands. It grows on rocky and sandy shores, often being common to abundant, and sometimes occurring inland at up to 150 m elevation in saline areas.

No uses are reported for this plant. It is very similar to another introduced species naturalized in Hawai'i, *Atriplex suberecta,* which differs in being an annual herb with smaller fruiting bracts that are yellow to brown at maturity; it grows in similar habitats, but sometimes occurs inland up to high elevations.

BACOPA MONNIERI
Scrophulariaceae (Snapdragon family)

English name: water hyssop
Vernacular names: none

Prostrate succulent herb often forming dense mats. Leaves simple, opposite, blade oblanceolate, 1–2.5 cm long, subsessile, surfaces glabrous. Flowers solitary in the leaf axils, with two linear bracts at the base. Calyx of 5 unequal, ovate, overlapping sepals 5–9 mm long. Corolla campanulate, lavender to white, deeply 5-lobed, 6–10 mm long. Stamens 4. Ovary superior, stigma capitate. Fruit an ovoid to conical capsule 5–8 mm long. SYNONYM: *Bramia monniera.*

Bacopa monnieri is pantropical in distribution, but in Polynesia is found only in the Marquesas and Hawai'i, and in Micronesia only in the Marianas (Guam) and Kiribati (formerly known as the Gilberts). It grows in areas of brackish water, such as coastal marshes, mudflats, and along estuaries, and is occasionally found inland in montane bogs (Marquesas).

No traditional uses have been reported for this plant in the Pacific islands, and since it is weedy and restricted to wet coastal areas, it is probably rarely even recognized locally. Nowadays, however, it is sometimes cultivated as a groundcover in Hawai'i. The Latin name has often been misspelled *Bacopa monniera.*

BOERHAVIA GLABRATA
Nyctaginaceae (Four-o'clock family)

Vernacular name: *alena* – Hawai'i

Slender prostrate herb forming low mats from a thick taproot. Leaves simple, opposite, blade lanceolate to narrowly ovate, 1–6 cm long, surfaces glabrous, petiole 5–20 mm long. Flowers several, in axillary umbels or compact cymes, on a thin peduncle 2–10 cm long. Calyx (perianth) petaloid, campanulate, white to pink, 1–3 mm long, shallowly 5-lobed. Corolla absent. Stamens usually 2, exserted. Ovary superior. Fruit an obovoid anthocarp 2–3.5 mm long, 5-ribbed, 1-seeded, sticky on the surface. SYNONYMS: *Boerhavia diffusa* of some authors, *Boerhavia acutifolia* of some authors.

Boerhavia glabrata is distributed from Java to Hawai'i, but in Polynesia and Micronesia appears to be restricted to Hawai'i (on all the main islands) and, apparently, French Polynesia. It grows in sunny, sandy or rocky, littoral habitats, rarely very far from the shore.

No uses are reported for this plant. It is very similar to *Boerhavia albiflora,* which is common on the central Pacific islands and rare on offshore islands of Australia and Samoa. It is also similar to *Boerhavia herbstii,* a rare endemic species of similar habitats in Hawai'i.

BOERHAVIA REPENS
Nyctaginaceae (Four-o'clock family)

Vernacular names: *alena* – Hawai'i; *katule* – Niue; *ufi'atuli* – Samoa; *akataha* – Tonga

Low slender herb with prostrate stems radiating from the thickened taproot. Leaves simple, opposite, blade narrowly lanceolate to oblong, mostly 1–4 cm long, tip mostly acute, surfaces glabrous, petiole 3–15 mm long. Flowers several, in axillary cymes, or nearly umbellate, on peduncles 2–10 cm long. Calyx (perianth) petaloid, campanulate, white to pink, 2–3 mm long, shallowly 5-lobed. Corolla absent. Stamens 2–4, exserted. Ovary superior. Fruit a club-shaped to ellipsoid anthocarp 3–4 mm long, 5-ribbed, 1-seeded, surface sticky. SYNONYM: *Boerhavia diffusa* of some authors.

Boerhavia repens is distributed from Africa to Hawai'i, and is found on all of the major archipelagoes of Polynesia and Micronesia. It is usually uncommon (but common in Hawai'i) in sandy or rocky littoral habitats, but is more frequently seen as a weed of coastal villages and plantations. Thus, it may be a weed introduced by the Polynesians rather than a native species.

The thickened root is reported to be a famine food in Samoa, Tonga, and other parts of Polynesia. In Hawai'i, a similar species, *Boerhavia coccinea,* is a common weed of littoral and coastal areas, and differs most noticeably in having ascending branches and red flowers.

BOERHAVIA TETRANDRA
Nyctaginaceae (Four-o'clock family)

Vernacular names: *nunanuna* (now forgotten?) – Societies; *moemoe* – Cooks (Pukapuka); *dafao* – Guam

Prostrate herb with stems up to 1 m long, radiating from a thick root. Leaves simple, opposite, blade ovate to suborbicular, 1 – 4 cm long, tip rounded, petiole 3 – 18 mm long. Flowers 3 – 10 in a simple or sometimes once-branching, axillary, subumbellate cluster on a peduncle up to 15 cm long. Calyx (perianth) campanulate, petaloid, lavender to pink, 2 – 3 mm long. Corolla absent. Stamens 2 – 4, exserted. Ovary superior. Fruit an ellipsoid anthocarp 2.5 – 3.5 mm long, sticky, 1-seeded. SYNONYM: *Boerhavia diffusa* var. *tetrandra.*

Boerhavia tetrandra is widely distributed in Polynesia and Micronesia (Marshalls and Kiribati), and is found on most of the atolls of the region. It usually grows in sandy, open littoral areas near the shore.

Few uses are reported for this plant, but in the Tuamotus and elsewhere, the root is reportedly cooked and eaten in times of famine. In Tokelau, the stems are sometimes fashioned into crude leis, and the leaves are sometimes used in native medicines.

CHAMAESYCE ATOTO
Euphorbiaceae (Spurge family)

Vernacular names: *'atoto* – Societies; *totototo* – Cooks; *toto* – Niue; *pulu tai*? – Samoa; *totolu, totoyava* – Fiji

Prostrate to erect, somewhat woody herb mostly less than 50 cm in height, with milky latex. Leaves simple, opposite, blade lanceolate to suborbicular, 6 – 50 mm long, surfaces glaucous, glabrous. Flowers in short, axillary cymes (cyathia), unisexual, plants monoecious. Involucre top-shaped, about 2 mm long, 4 – 5-lobed, with 4 – 5 white to green glands. Corolla absent. Stamens in male flowers 10 – 15. Ovary in female flowers superior, style 3-lobed. Fruit a subglobose, 3-lobed capsule about 2 mm long, splitting into three 1-seeded segments. SYNONYMS: *Euphorbia ramosissimum, Euphorbia atoto, Euphorbia tahitensis, Euphorbia chamissonis* of some authors, *Chamaesyce chamissonis* of some authors.

Chamaesyce atoto is distributed from Ceylon eastward to the Marquesas in Polynesia. Some botanists consider this "species" to actually consist of several related species, such as *Chamaesyce chamissonis* of Micronesia. These plants are found on all the high archipelagoes of Polynesia (except Hawai'i) and all the major archipelagoes of Micronesia, growing on sandy or rocky shores and sometimes inland in sunny places in the mountains and on foothill ridges and cliffs.

The plant is occasionally employed in native medicines in the Tuamotus. In the Society Islands, a brown dye for tinting garments was reportedly made from its roots. In the Cook Islands, Tonga, and elsewhere, children write on their skin with the sap, allow it to dry clear, and then rub dust or ashes across it to make the writing appear "magically."

CHAMAESYCE DEGENERI
Euphorbiaceae (Spurge family)

Vernacular name: *'akoko* – Hawai'i

Woody herb or subshrub with prostrate to ascending stems up to 40 cm long, with milky latex. Leaves simple, opposite, orbicular to broadly elliptic, 1 – 1.5 cm long, subsessile, surfaces mostly glabrous. Flowers in solitary cymes (cyathia) in the upper axils, unisexual, plants monoecious. Involucre campanulate, 1 – 2 mm long, with 4 green glands. Corolla absent. Stamens of male flowers 10 – 15. Ovary of female flowers superior, 3-celled, with three 2-lobed stigmas. Fruit a subglobose capsule 1.5 – 2.5 mm long, splitting into three 1-seeded segments. SYNONYM: *Euphorbia degeneri*.

Chamaesyce degeneri is endemic to Hawai'i, where it is reported from all the major islands except Lanai and Kaho'olawe. It grows on sand dunes and rocky coastal slopes, but is uncommon and restricted to scattered localities.

No uses are reported for this plant, and its common name applies to many members of the genus, probably because the white sap that exudes from the cut stems and leaves is reminiscent of blood (*koko*). There are about 14 other endemic species of this genus in Hawai'i, but only one, *Chamaesyce skottsbergii*, could also be considered littoral.

CRESSA TRUXILLENSIS
Convolvulaceae (Morning-glory family)

Vernacular names: none

Small herb with procumbent to erect stems up to 35 cm long. Leaves simple, alternate, blade fleshy, elliptic to lanceolate, 3–10 mm long, subsessile, surfaces and stems gray, pubescent. Flowers solitary, axillary, on a pedicel 1–5 mm long. Calyx of 5 unequal, elliptic to ovate sepals 3–5 mm long. Corolla salverform, white, throat 5–7 mm long, limb divided about halfway into 5 lobes. Stamens 5, exserted. Ovary superior, styles 2, each 2-lobed. Fruit an ovoid capsule 5–6 mm long, usually containing 2 glabrous, ovoid seeds 3–4 mm long. SYNONYMS: *Cressa insularis, Cressa cretica* of some authors.

Cressa truxillensis, formerly called *Cressa insularis* in Hawai'i, was previously thought to be endemic there, but is also distributed from the southwestern U. S. to South America. In Hawai'i, it is reported from O'ahu (where it may now be extinct), Moloka'i, and Kaho'olawe. It grows on mudflats and on the inland side of sandy beaches, most notably now on the west end of Moloka'i.

No uses or Hawaiian names have been reported for this plant, probably because it is uncommon and inconspicuous. Its rarity in Hawai'i and probable extinction from O'ahu may be due to competition in its prime habitats from introduced littoral plants such as *Atriplex semibaccata*.

GNAPHALIUM SANDWICENSIUM
Asteraceae (Sunflower family)

Vernacular name: *'ena'ena* – Hawai'i

Perennial herb with woolly, erect to prostrate stems 10–60 cm long. Leaves simple, alternate, spathulate to narrowly oblanceolate, mostly 1–7 cm long, sessile, surfaces densely woolly. Flowers in terminal, subglobose clusters of heads, each 2–3 mm long and surrounded by many membranous, overlapping bracts. Ray florets absent. Disc florets 2–3 mm long, bisexual or female, yellow, 5-lobed. Fruit an oblong, brown achene less than 1 mm long. SYNONYM: *Gnaphalium hawaiiense*.

Gnaphalium sandwicensium is endemic to Hawai'i, where it is found on all the main islands and on the atolls of the Leeward Islands. It grows in dry places such as coastal sand dunes near sea level, and also inland on cinder or lava at up to 3000 m elevation, making it a facultative rather than an obligate littoral species.

The plant emits a balsam-like fragrance and ancient Hawaiians reportedly placed it in calabashes as an insect repellant for feather cloaks. Several varieties are recognized. This native littoral species is easily confused with a weedy introduced species, *Gnaphalium purpureum,* an annual herb with heads in spike-like inflorescences that grows in disturbed places in Hawai'i and Tonga.

HALORAGIS PROSTRATA
Haloragaceae (Water milfoil family)

Vernacular names: none

Low woody herb or subshrub with erect to ascending stems up to 30 cm or more in height. Leaves simple, opposite and decussate, blade narrowly elliptic to oblanceolate, 1.5–4.5 cm long, subsessile, surfaces glabrous. Flowers several in short, sessile, axillary clusters, unisexual, plants monoecious. Calyx of 4 ovate sepals about 1.5 mm long. Corolla of 4 white petals 2–4 mm long, enfolding the anthers. Stamens 4 or 8, probably vestigial in female flowers. Ovary inferior, stigma 4-lobed, plumose, probably vestigial in male flowers. Fruit a globose, 4-seeded drupe 2–3 mm long, reddish at maturity.

Haloragis prostrata is found in New Caledonia, Vanuatu, and far to the east on the tiny island of Miti'aro in the Cook Islands. On Miti'aro, where it was only recently discovered, it is quite common growing in the low scrubby vegetation on makatea rock along the shores.

No uses or native names are reported for the plant. Although Miti'aro is very small, very low, and mostly covered by lakes and makatea, several other interesting plants occur there, such as the only sandalwood species in the Cook Islands, and *Tetramolopium sylvae*, which was originally thought to be endemic to Hawai'i.

HEDYOTIS BIFLORA
Rubiaceae (Coffee family)

Vernacular names: none

Erect, diffusely branching herb up to 40 cm in height, with interpetiolar stipules. Leaves simple, opposite, elliptic to ovate, 1 – 6 cm long, base attenuate, surfaces glabrous, petiole up to 1 cm long. Flowers mostly 3 – 7, in delicate, terminal or axillary panicles 1 – 8 cm long. Calyx urn-shaped, about 1 – 2 mm long, divided into 4 triangular, acute lobes. Corolla broadly tubular, white, 2.5 – 4 mm long, divided less than halfway into 4 acute lobes. Stamens 4, epipetalous. Ovary inferior, style 2-lobed. Fruit a subglobose capsule 2.5 – 4 mm long with the persistent calyx lobes on top, containing numerous tiny black seeds. SYNONYMS: *Hedyotis paniculata, Oldenlandia paniculata, Oldenlandia biflora.*

Hedyotis biflora is distributed from the Indian Ocean to the Cook Islands (Miti'aro). In Polynesia, it is found in Fiji, Samoa, Tonga, Niue, and the Cook Islands; in Micronesia, it occurs on all the major archipelagoes. It grows in shallow soil in littoral forest, or more commonly in cracks and crevices on rocky coasts, but sometimes inland in streambeds at low elevations, and occasionally as a weed.

Since the plant is so inconspicuous and uncommon, no names or native uses have been reported for it in the islands. A related species, *Hedyotis foetida*, differs in being somewhat woody with larger leaves, and has more numerous and larger flowers.

HEDYOTIS FOETIDA
Rubiaceae (Coffee family)

Vernacular name: *kinakina*? – Cooks

Somewhat woody herb up to 50 cm in height, with 4-angled stems and interpetiolar stipules. Leaves simple, opposite, blade elliptic, 3–9 cm long, surfaces glabrous, petiole up to 1 cm long. Flowers many, in terminal and upper-axillary panicles 2–12 cm long. Calyx urn-shaped, 1.5–3 mm long, divided about halfway into 4 lobes, pedicel 1–5 mm long. Corolla funnelform, white, 4–8 mm long, divided less than halfway into 4 acute lobes. Stamens 4, epipetalous. Ovary inferior. Fruit an obovoid capsule 2–3 mm long with the persistent calyx lobes on top, containing numerous tiny, dull brown seeds.

Hedyotis foetida is distributed from Melanesia and Micronesia (the Marianas) to eastern Polynesia (the Australs), and is found on most of the high islands in the area. It usually grows on rocky shores between the littoral forest and the beach, but sometimes can be found in sunny, rocky places in the mountains (Rarotonga and Tonga). A similar, related species, *Hedyotis biflora*, differs most obviously in having shorter leaves and flowers. Another littoral species, *Hedyotis romanzoffiensis*, differs in having fleshy fruits rather than a dry capsule.

The plant is not well known over most of its range, and no vernacular names or significant uses have been reported for it.

HEDYOTIS ROMANZOFFIENSIS
Rubiaceae (Coffee family)

Vernacular names: *kōporoporo, poroporo* – Tuamotus; *polo* – Cooks (Pukapuka); *kautokiaveka* – Tokelau

Small woody herb or subshrub 30 – 200 cm in height, with interpetiolar stipules. Leaves simple, opposite, blade obovate, 2 – 7 cm long, surfaces glabrous. Flowers solitary or 2 – 3 in axillary cymes 5 – 30 mm long. Calyx urn-shaped, 3 – 5 mm long, 4 – 5-lobed at the tip, pedicel 5 – 15 mm long. Corolla salverform, white, tube about 5 mm long, with 4 ovate lobes 4 – 5 mm long. Stamens 4, epipetalous. Ovary inferior, style 2-lobed. Fruit an obovoid, fleshy, drupe-like berry 10 – 15 mm long, white turning purple at maturity.

Hedyotis romanzoffiensis is endemic to Polynesia, where it occurs in the Society Islands (Tupa'i and Scilly), northern Cook Islands, Australs, Tuamotus, Line Islands, Tokelau, and Tuvalu. It is restricted to sunny littoral habits on sandy beaches of atolls, and only rarely occurs on high islands (the Australs). It differs from most other Polynesian species of *Hedyotis* in having a fleshy berry rather than a dry capsule.

The fruit is reportedly eaten by pigeons, and is sometimes strung into leis used for decoration, but few other significant uses have been reported for the plant.

HELIOTROPIUM ANOMALUM
Boraginaceae (Borage family)

Vernacular names: *hinahina* – Hawai'i; *toihune fifine* – Niue

Prostrate herb up to 30 cm in height, with stems and foliage densely covered with silky, appressed hairs. Leaves alternate or appearing densely whorled, blade linear-lanceolate to spathulate, 1–5 cm long. Flowers many, in short, congested axillary cymes on a peduncle 2–6 cm long. Calyx deeply divided into 5 linear sepals 2–4 mm long. Corolla funnelform, white with a yellow center, fragrant, 6–11 mm long, limb 5–6-lobed. Stamens 5 or 6. Ovary superior, deeply 4-lobed. Fruit consisting of 4 obovoid nutlets about 1 mm long.

Heliotropium anomalum is widely distributed from the Ryukyu Islands to eastern Polynesia. In Polynesia, it is found in Hawai'i, the Tuamotus, Society Islands (Tahiti, but apparently rare there), Australs, and Niue; in Micronesia, it occurs in the Marianas, Carolines, and Wake Island. It is a common species on limestone shores and sandy beaches throughout it range.

Although the plant is common in some places, it often goes nameless in the Pacific islands. It has few reported uses, but its fragrant flowers and leaves are sometimes used for making leis in Hawai'i.

HELIOTROPIUM CURASSAVICUM
Boraginaceae (Borage family)

English name: seaside heliotrope
Vernacular names: *nena, kīpūkai* – Hawai'i

Prostrate to decumbent, perennial herb with stems up to 45 cm long. Leaves simple, alternate, blade oblanceolate, 2–5 cm long, subsessile, surfaces glabrous. Flowers many, in terminal or axillary, single or paired, scorpeoid cymes mostly 2–5 cm long. Calyx divided to near base into 5 lanceolate to oblong lobes 1–2 mm long. Corolla funnelform, white with a green or yellow center, 1–4 mm long, limb 5-lobed. Stamens 5. Ovary superior. Fruit consisting of 4 ovoid, 1-seeded nutlets 1.5–2 mm long.

Heliotropium curassavicum is distributed from the southern United States to Australia, but in the Pacific islands is found only in Hawai'i, where it is reported from all the main islands. It is occasional to common in salt flats, marshy areas, and rocky beaches or coastal slopes in open vegetation, rarely growing very far inland.

It is similar to, but not nearly as attractive as, *Heliotropium anomalum,* which is found in Hawai'i and some of the other Pacific islands; no significant uses are reported for it. A similar species, *Heliotropium procumbens,* is found as a weed in Micronesia (Marianas, Carolines, Marshalls) and Hawai'i; it has pubescent stems and bracts on the inflorescence.

LEPIDIUM BIDENTATUM
Brassicaceae (Mustard family)

English name: scurvy grass
Vernacular names: *'ānaunau* – Hawai'i; *nau, horahora* – Societies; *nau, naunau* – Cooks

Erect perennial herb up to 60 cm in height, somewhat woody at the base. Leaves simple, alternate, blade somewhat fleshy, obovate to spathulate, 4 – 10 cm long, margins toothed, surfaces glabrous. Flowers many, in terminal racemes up to 30 cm long. Calyx of 4 ovate sepals 1 – 1.5 mm long. Corolla of 4 white petals 1 – 2 mm long. Stamens 6. Ovary superior. Fruit a flattened, ellipsoid to subglobose capsule 4 – 8 mm long, on a pedicel of similar length. SYNONYMS: *Lepidium piscidium, Lepidium o-waihiense, Lepidium bidentoides.*

Lepidium bidentatum is distributed from New Caledonia to eastern Polynesia, occurring in Hawai'i, the Tuamotus, Society, Austral, and Cook Islands, but absent from Fiji, western Polynesia, and Micronesia (except Wake Island). It grows on limestone, basaltic rock, or coral sand, but usually not far from the sea, and is particularly common on atolls.

The edible leaves are eaten raw or cooked in the Society Islands and Tuamotus, and during the early European period in Polynesia, they were collected and fed to sailors to prevent scurvy. They are also sometimes used in native medicines in the Society Islands and Cook Islands.

LIPOCHAETA INTEGRIFOLIA
Asteraceae (Sunflower family)

Vernacular name: *nehe* – Hawai'i

Low perennial herb with prostrate stems up to 2 m long, forming roots along their lower surface. Leaves simple, opposite, blade thick and somewhat fleshy, oblong to spathulate, mostly 5–20 mm long, margins entire to slightly serrate, surfaces appressed-hairy, petiole short. Flowers in globose, long-stalked composite heads 5–15 mm in diameter. Ray florets ovate to oblong, 3–5 mm long, yellow, 4–10 per head; disc florets 25–50 per head, with a yellow tubular corolla 2.5–3.5 mm long. Fruit an awnless achene 2–3 mm long.

Lipochaeta integrifolia is endemic to Hawai'i, where it is found on all the main islands. It grows in scattered localities on rocky slopes and sandy shores, sometimes being locally common, but rarely occurring very far from the shore.

No uses have been reported for this plant, but it can be an important part of the vegetation in native habitats where it occurs. There are twenty endemic species of *Lipochaeta* in Hawai'i, and a few of these, both living and extinct, have been reported to be growing in or near littoral habitats, but *Lipochaeta integrifolia* is the species most likely to be seen on the shore. The genus is closely related to *Wedelia,* the genus to which *Wollastonia biflora* was formerly assigned.

LYSIMACHIA MAURITIANA
Primulaceae (Primula family)

Vernacular names: none

Perennial herb with stems up to 80 cm long, becoming woody at the base. Leaves simple, alternate, blade somewhat fleshy, narrowly oblanceolate to elliptic, 1–6 cm long, sessile or on a short petiole, surfaces glabrous with scattered black dots below. Flowers many, in dense, leafy, terminal racemes 2–15 cm long. Calyx divided to the base into 5 lanceolate lobes 4–6 mm long, on a pedicel 3–20 mm long. Corolla campanulate, white or tinged pink, 8–14 mm long, deeply 5-lobed. Stamens 5. Ovary superior. Fruit an ovoid or pear-shaped capsule 4–6 mm long, many-seeded, with a persistent, unbranched style 2–3 mm long.

Lysimachia mauritiana is distributed from East Africa to the Pacific islands. Although widely found in the western Pacific, in Micronesia it is recorded only from the Marianas, and only from Hawai'i (Ni'ihau, Kaua'i, Moloka'i, Maui, and Hawai'i) in Polynesia. It grows on coastal rocks, sea cliffs, and rocky beaches, but is uncommon and restricted to scattered localities.

No uses or Hawaiian names are reported for this species, probably because of its infrequent occurrence and inconspicuous nature. In addition to this indigenous species, there are about ten endemic species of *Lysimachia* in Hawai'i, none of which are littoral. A related species, *Lysimachia rapensis,* is reported from Rapa.

NAMA SANDWICENSIS
Hydrophyllaceae (Waterleaf family)

Vernacular name: *hinahina kahakai* – Hawai'i (Ni'ihau)

Perennial prostrate herb forming mats up to 20 cm in diameter, with branching, densely pubescent stems. Leaves simple, alternate, blade thick, spathulate, mostly 6 – 20 mm long, margins revolute, surfaces covered with appressed hairs. Flowers 1 – 2 in leaf axils. Calyx 4 – 7 mm long, deeply divided into 5 narrow lobes. Corolla narrowly campanulate, purplish blue to pink or white, 5 – 8 mm long, shallowly divided into 5 rounded lobes. Stamens 5. Ovary superior, stigma 2-lobed to the base. Fruit an oblong to ovoid capsule 3 – 4 mm long, containing 20 – 60 tiny seeds.

Nama sandwicensis is endemic to Hawai'i, where it is found on all the main islands. It is uncommon and inconspicuous on sandy beaches and on raised limestone coasts, rarely occurring very far inland. Except for this Hawaiian endemic species, the other 50 or so species of the genus are native from the southwestern U. S. to South America.

No uses are reported for this plant, and because it is so small, infrequent, and inconspicuous, it is easily overlooked, so that few people are even aware of its existence.

NESOGENES EUPHRASIOIDES
Chloanthaceae (Chloanthus family)

Vernacular names: none

Prostrate to ascending, somewhat woody herb up to 60 cm in height, with many stems radiating from the rootstock. Leaves simple, opposite, blade somewhat fleshy, ovate to narrowly elliptic, 6–20 mm long, surfaces mostly covered with short, stiff hairs. Flowers 1 or 2 in the axils, on pedicels about 1 mm long. Calyx campanulate, 2–4 mm long, divided less than halfway into 5 triangular lobes. Corolla campanulate, white mottled with red, 4–6 mm long, 5-lobed. Stamens 4. Ovary superior. Fruit a campanulate capsule 3–5 mm long (including the persistent calyx lobes), 1- or 2-seeded.

Nesogenes euphrasioides is endemic to eastern Polynesia, where it is found in the Tuamotus, Society Islands (Tupai), and Cook Islands (Ma'uke, Miti'aro). It grows on makatea, coral sand, or coral rubble of atolls and makatea islands, and occasionally inland in littoral forest. The genus, which consists of only 2 or 3 species, has an odd distribution; the other species occur thousands of miles away in the Indian Ocean (a third species may occur in the Marianas).

The plant is inconspicuous and is rarely recognized or named by the inhabitants of the islands where it occurs. No significant uses have been reported for it.

NICOTIANA FRAGRANS
Solanaceae (Nightshade family)

Vernacular names: none

Low perennial herb up to 50 cm in height, base somewhat woody. Leaves simple, alternate, oblanceolate to spathulate, 3–12 cm long, attenuate on a short petiole, surfaces viscous. Flowers several, in axillary racemes or narrow panicles up to 25 cm long. Calyx campanulate, 9–12 mm long, divided about halfway into lanceolate sticky lobes. Corolla salverform, white, sticky on the outside, tube 3.5–4.5 cm long, limb about 15–20 mm wide, divided into 5 rounded lobes. Stamens 5. Ovary superior. Fruit an ovoid, many-seeded capsule up to 1 cm long, enclosed within the persistent calyx.

Nicotiana fragrans is found only in New Caledonia (Isle of Pines), Tonga (Tongatapu), and Niue. It grows on coastal limestone rocks and cliffs; it is rare in Tonga, but more frequent on Niue. A related species endemic to the Marquesas, *Nicotiana fatuhivensis*, grows on sea cliffs on most of the islands, and inland up to 800 m in elevation. It differs from *Nicotiana fragrans* in having elliptic to obovate leaves, a broader panicle with more numerous flowers, a larger calyx (12–16 mm long), and a longer corolla (4–7 cm long) with a broader limb (4 cm and larger).

Because of the rarity of these two species and their inaccessible habitat, the plants are not well known where they occur and have no reported local names or uses.

PORTULACA LUTEA
Portulacaceae (Purslane family)

Vernacular names: *'ihi* – Hawai'i; *'aturi* – Societies; *pokea* – Cooks; *kamole* – Niue; *tamole* – Samoa, Tonga

Prostrate to ascending, succulent, perennial herb with a swollen tuberous root. Leaves simple, mostly alternate, blade ovate to suborbicular, mostly 5–30 mm long, surfaces glabrous. Flowers 1–3, in terminal, congested cymes. Calyx of 2 suborbicular sepals 7–9 mm long. Corolla of 5 yellow obovate petals 9–12 mm long. Stamens 18–50, yellow. Ovary superior. Fruit an ovoid, circumscissile capsule 6–8 mm long, opening by means of a cap that splits off to release the numerous, tiny, shiny black seeds. SYNONYM: *Portulaca johnii.*

Portulaca lutea is distributed from New Caledonia to Hawai'i, and is found in most of the major Polynesian archipelagoes (except the Tuamotus), and in the Marianas (but not Guam), Marshalls, and Kiribati (formerly known as the Gilberts) of Micronesia. It grows on lava or limestone shores and sandy beaches, often as a dominant species. Unlike the related cosmopolitan weed, *Portulaca oleracea*, it is rarely found far inland and is restricted to native habitats. Another related littoral species, *Portulaca molokiniensis*, is endemic to Hawai'i, but is very rare.

The enlarged, edible root was once cooked with coconut cream, but nowadays serves mostly as food for pigs. The plant is also sometimes an ingredient in native medicines in the Society Islands.

PORTULACA SAMOENSIS
Portulacaceae (Purslane family)

Vernacular name: *tamole* – Samoa

Perennial succulent herb with prostrate stems up to 20 cm or more in length, bearing tufts of hairs up to 5 mm long in the leaf axils. Leaves simple, alternate or opposite, blades narrowly elliptic to lanceolate, 5–10 mm long, subsessile, surfaces glabrous. Flowers 2–8, in a short terminal cluster. Calyx of 2 sepals 3–4.5 mm long. Corolla of 4 or 5 yellow, obovate petals mostly 4–8 mm long. Stamens 20–30, yellow. Ovary inferior. Fruit a subglobose, circumscissile capsule 2.5–4.5 mm long, opening by means of a cap that splits off to expose the numerous, tiny, dark-gray seeds. SYNONYMS: *Portulaca pilosa* of some authors, *Portulaca australis* of some authors.

Portulaca samoensis is distributed from New Caledonia to western Polynesia, occurring on all the major archipelagoes of Micronesia, and in Samoa, Tonga, Niue, and Tokelau of western Polynesia. It grows on sandy beaches and in crevices on rocky coasts and slopes up to 200 m elevation. Some botanists believe this to be the same as an Australian species, *Portulaca australis*; if so, the latter name would be the correct one.

Other than its occasional inclusion in native medicines in Fiji, no significant uses are reported for this species. It is much smaller and less conspicuous than the other littoral species of *Portulaca* included here.

PORTULACA VILLOSA
Portulacaceae (Purslane family)

Vernacular name: *'ihi* – Hawai'i

Perennial succulent herb with prostrate to erect stems up to 30 cm long from a woody taproot, and bearing a dense tuft of hairs 3–12 mm long in the axils. Leaves simple, alternate, linear to narrowly oblanceolate, 5–15 mm long, sessile, tip obtuse, base attenuate, surfaces glabrous. Flowers in terminal clusters of 3–6, with a dense tuft of hairs at their base. Calyx of 2 sepals 4–5 mm long. Corolla of 5 obovate, white to pink petals notched at the tip, 8–10 mm long. Stamens 18–50, yellow. Ovary superior. Fruit an ovoid, circumscissile capsule 3–5 mm long, opening by means of a cap that splits off to release the numerous tiny, dark-brown seeds. SYNONYMS: *Portulaca pilosa* of some authors, *Portulaca hawaiiensis.*

Portulaca villosa is endemic to Hawai'i, where it has been recorded from all the main islands except Kaua'i and Ni'ihau. It is rare, however, and is now restricted to a few scattered localities, and is perhaps most common on the south and southeast coast of the island of Hawai'i. It grows on rocky lava or coralline coasts, and occasionally inland up to 300 m in elevation.

No uses have been recorded for this species. It has previously been confused with a naturalized species, *Portulaca pilosa* (also known as *Portulaca cyanosperma*); this species has magenta flowers and is occasional to common in coastal and inland habitats.

SCHIEDEA GLOBOSA
Caryophyllaceae (Carnation family)

Vernacular names: none

Low woody herb or subshrub with prostrate to ascending stems up to 40 cm long, forming clumps from the woody base. Leaves simple, opposite, blade somewhat fleshy, linear to oblanceolate, mostly 3–10 cm long, subsessile, surfaces glabrous. Flowers many, in a dense, terminal, globose cyme 15–40 mm in diameter, often with a smaller pair of clusters at the next highest node, mostly unisexual, plants dioecious. Calyx of 4–6 ovate sepals 3–5 mm long, green. Corolla absent. Stamens 10, exserted, or reduced to staminodes in female flowers. Ovary superior, with 4 styles, vestigial in male flowers. Fruit an ovoid capsule 2–4.5 mm long, containing several seeds.

Schiedea globosa is endemic to Hawai'i, where it is reported from O'ahu, Moloka'i, and Maui. It grows on rocky coastal slopes and sea cliffs, from near sea level to 300 m elevation, but is nowhere reported to be common.

Because it is uncommon and inconspicuous, no uses or Hawaiian names have been noted for this species. The genus *Schiedea* is endemic to Hawai'i and consists of about 22 species, but *Schiedea globosa* is the only one that can be considered littoral.

SESUVIUM PORTULACASTRUM
Aizoaceae (Iceplant family)

English name: sea purslane
Vernacular name: *'ākulikuli* – Hawai'i

Prostrate succulent herb with red stems up to 80 cm long, often forming dense mats. Leaves simple, opposite, blade succulent, linear to spathulate, 1 – 3.5 cm long, surfaces glabrous. Flowers solitary, axillary on a short pedicel up to 12 mm long. Calyx 6 – 9 mm long, petoloid and white to pink inside, deeply divided into 5 acute lobes. Corolla absent. Stamens 5 to many. Ovary superior, styles 3 – 5. Fruit an ovoid to subglobose capsule 5 – 7 mm long, containing numerous small black seeds.

Sesuvium portulacastrum is pantropical in distribution, and is found on all the high archipelagoes of Polynesia (although not recorded from the Society Islands in many years), and in the Marianas, Marshalls, and Kiribati (formerly known as the Gilberts) of Micronesia. It grows on rocky shores, sea cliffs, and in salt marshes, and is one of the most salt-tolerant of the littoral species; it grows in and dominates wave-splashed or saline habitats, where few other species can survive.

This herb is not well known to the Polynesians, and no significant uses have been reported for it. It is easily confused with the similar *Portulaca lutea*, which differs, however, in having yellow flowers and wider, shorter leaves.

SOLANUM AMICORUM
Solanaceae (Nightshade family)

Vernacular name: *polo Tonga* – Tonga

Woody herb or subshrub up to 1 m in height, with young stems covered by stellate hairs. Leaves simple, alternate, blade mostly ovate, 10–20 cm long, margins entire to wavy, upper surface nearly glabrous, lower surface covered with stellate hairs, especially on the veins, petioles 3–5 cm long. Flowers several, borne on the stem in short racemes or panicles 1–4 cm long. Calyx campanulate, 2.5–3.5 mm long, divided about halfway into 5 linear-lanceolate lobes, on a thin pedicel mostly 8–14 mm long. Corolla rotate, white, 7–12 mm long, divided about halfway into 5 triangular lobes. Stamens 5, yellow, subsessile. Ovary superior. Fruit a red, globose berry 6–10 mm in diameter, many-seeded.

Solanum amicorum is found only in Tonga (mostly in the northern islands) and the nearby island of Niue, but there is also a questionable, century-old record of it from Samoa. It grows in sunny places in coastal thickets and in openings in littoral forest, but is rare over most of its range.

No uses are reported for this plant, and its Tongan name is also usually applied to the somewhat more common *Solanum viride*. This latter species, which is native or an early Polynesian introduction, is sometimes used in native medicines, and is found throughout Polynesia in coastal and lowland habitats on high islands and atolls.

SOLANUM NELSONII
Solanaceae (Nightshade family)

Vernacular name: *pōpolo* – Hawai'i

Trailing or prostrate woody herb or subshrub up to 1 m in height, with stems and leaves densely covered with golden, stellate hairs. Leaves simple, alternate, blade ovate to suborbicular, 1 – 10 cm long, base rounded to cordate or oblique on a petiole less than half as long as the blade, margins entire or shallowly cut into round lobes. Flowers 1 – several, in racemose cymes on a peduncle up to 5 cm long. Calyx 3 – 5 mm long, shallowly divided into 5 triangular lobes, densely covered with stellate hairs on the outside. Corolla rotate, purple to white, 6 – 12 mm long, deeply divided into 5 lobes, purple to white. Stamens 5. Ovary superior. Fruit a globose berry 4 – 10 mm in diameter, black at maturity, containing numerous seeds. SYNONYM: *Solanum laysanense.*

Solanum nelsonii is endemic to Hawai'i, where it is found on the Leeward Islands and on all the main islands except Lana'i and Kaho'olawe. Although common in the Leewards, it is rare on the main islands, where it occurs only in scattered localities, most notably on Moloka'i. The plant grows on sandy or coral rubble coasts, and rarely inland up to 150 m in elevation.

No uses are reported for this plant. It is related to another endemic Hawaiian species, *Solanum incompletum*, but this shrub is an inland species. The name *pōpolo* is a general one, and more commonly applies to a weedy species, *Solanum americanum* (formerly known as *Solanum nigrum*), used in native medicines.

TACCA LEONTOPETALOIDES
Taccaceae (Polynesian arrowroot family)

English name: Polynesian arrowroot
Vernacular names: *pia* – Hawai'i, Societies, Cooks, Niue; *māsoā* Samoa; *mahoa'a* – Tonga; *yabia* – Fiji; *gapgap* – Guam

Stemless herb growing from a starchy tuber. Leaves 1 – 3, irregularly divided into 3 parts, deeply lobed, erect, up to 1 m in height, glabrous. Flowers 20 – 40 in an umbel borne atop a hollow, grooved scape up to 1.3 m or more in height. Umbels with 4 – 12 lanceolate to ovate bracts 3 – 6 cm long, and numerous filamentous bracts 10 – 20 cm long. Perianth campanulate, mostly 6 – 12 mm long, with 6 green tepals 4 – 8 mm long. Fruit a subglobose berry 1.5 – 2.5 cm or more in diameter, green, many-seeded. SYNONYM: *Tacca pinnatifida.*

Tacca leontopetaloides is distributed from India to Hawai'i, and is found on all the major archipelagoes of Polynesia and Micronesia. It is probably an ancient introduction to Polynesia, but is naturalized in littoral thickets and forests on sandy shores of high islands and atolls, but only occasionally is it found very far inland.

Starch obtained from the grated tuber was once a major source of food, particularly when mixed as a thickener in other foods. The starch was also used to stiffen fabrics, and in Hawai'i and other places, was employed in native medicines. In the Marquesas, the long flowering stalk was cut into short sections and strung into leis, and in the Society and Cook Islands, the stem fibers were used in weaving.

TEPHROSIA PURPUREA
Fabaceae (Pea family)

Vernacular names: *'auhuhu, 'auhola, hola*—Hawai'i; *hora*—Societies; *mata'ora*—Cooks; *kohuhu*—Niue; *kavahuhu*—Tonga; *'avasā*—Samoa; *tuvakei*—Fiji

Annual or perennial, somewhat woody herb up to 1 m or more in height. Leaves alternate, odd-pinnately compound, rachis mostly 3–8 cm long, leaflets 7–15, blades mostly oblanceolate, 8–32 mm long, appressed-hairy on lower surface, tip rounded or notched. Flowers several, in terminal or leaf-opposed racemes up to 15 cm or more long. Calyx 3–4 mm long, divided about halfway into 5 lobes. Corolla papilionaceous, white or occasionally purple, 6–9 mm long. Stamens 10, diadelphous. Ovary superior. Fruit a linear pod 2–5 cm long, twisting apart at maturity, usually with 5–9 seeds 2.5–5 mm long. SYNONYM: *Tephrosia piscatoria.*

Tephrosia purpurea is distributed from East Africa to eastern Polynesia, and is found on most of the high islands in this area. It may, however, be an ancient introduction to Polynesia, at least to the eastern portions, as it certainly is in Hawai'i. It grows on rocky shores and sometimes inland in rocky, sunny places, but nowadays is becoming rare in Polynesia. Although it is absent from Micronesia, another related species, *Tephrosia mariana,* occurs there.

The only major use for this plant has been as a fish poison. Throughout Polynesia, the roots and foliage were crushed and thrown into the lagoon. The poison stuns the fish, which float to the surface and are collected; it affects fish, but not humans.

TETRAGONIA TETRAGONIOIDES
Aizoaceae (Iceplant family)

English name: New Zealand spinach
Vernacular names: none

Spreading or prostrate, succulent herb with stems up to 1.5 m long. Leaves simple, alternate, blade triangular to subcordate, mostly 2–10 cm long, surfaces rough, petiole winged. Flowers solitary or paired in the axils. Calyx of 4 or 5 spreading sepals 1.5–2.5 mm long, green on the outside, yellow within, pedicel 2–5 mm long. Corolla absent. Stamens 10–20. Ovary inferior. Fruit top-shaped, 4-angled, 8–12 mm long with 2–5 horn-like projections on the top, 3–8-seeded. SYNONYM: *Tetragonia expansa*.

Tetragonia tetragonioides is distributed around the Pacific in South America, New Zealand, Australia, and Japan, but in Polynesia appears to be indigenous only to Tonga and the Austral Islands. It was introduced to Hawai'i, however, where it is naturalized in scattered localities and is occasionally cultivated. It grows mostly in sandy or rocky littoral areas, but nowhere is it reported to be common.

The plant is eaten as a vegetable, and is sometimes cultivated for this purpose. No names are reported for it in its native Polynesian range, and it is probably infrequently noticed by local inhabitants.

TETRAMOLOPIUM ROCKII
Asteraceae (Sunflower family)

Vernacular names: none

Prostrate perennial herb 5 – 10 cm in height, forming low, compact mats. Leaves simple, alternate and forming basal rosettes, blade spathulate, 15 – 30 mm long including the attenuate petiole, surfaces glandular, finely pubescent. Flowers in solitary composite heads 1 – 1.7 cm in diameter, subtended by numerous lanceolate bracts in a single series, on a scape 4 – 12 cm long. Ray florets mostly 60 – 100 per head, white, 3 – 4.5 mm long; disc florets mostly 30 – 55 per head, yellow, 4 – 5 mm long. Fruit an oblanceoloid to cylindrical achene 2 – 3 mm long, with a terminal pappus of white bristles about as long.

Tetramolopium rockii is endemic to Hawai'i, where it is known only from the fossil sand dunes of Mo'omomi beach on the north coast of Moloka'i. A closely related species, *Tetramolopium sylvae*, is found on seacliffs of Maui and the north coast of Moloka'i, as well as on coralline beach deposits of Miti'aro in the Cook Islands.

No uses or Polynesian names have been reported for this species, probably because it is so rare and inconspicuous. In addition to the two species mentioned here, there are nine other endemic species of *Tetramolopium* in Hawai'i, and several in New Guinea, but none of these grow in littoral habitats.

TRIBULUS CISTOIDES
Zygophyllaceae (Caltrop family)

Vernacular name: ***nohu*** – Hawai'i

Prostrate perennial herb with densely pubescent stems. Leaves opposite, pinnately compound, rachis mostly 5–12 cm long, leaflets opposite, blades oblong, 8–24 mm long, sessile, silvery pubescent. Flowers solitary, on a long, axillary pedicel. Calyx 6–10 mm long, split to the base into 5 lanceolate lobes. Corolla of 5 obovate, yellow petals 15–20 mm long. Stamens 10. Ovary superior, 5-lobed. Fruit a green, spiny schizocarp of 5 sections, each of which bears two spines up to 8 mm long.

Tribulus cistoides is pantropical in distribution, and is widespread in Polynesia (Marquesas, Tuamotus, northern Cooks, Hawai'i), and in Micronesia (Marianas, Marshalls, Kiribati). It usually grows on sandy shores, but is occasionally found inland in open places at up to 400 m elevation (Marquesas).

The plant has no reported uses, and is even a nuisance because of the sharp spines on the fruit. It is closely related to the puncture vine, *Tribulus terrestris*, a widespread noxious weed of temperate areas of the world.

TRIUMFETTA PROCUMBENS
Tiliaceae (Linden family)

Vernacular names: *'urio* – Societies; *vavai?* – northern Cooks; *sisi tai* – Niue; *masiksik hembra* – Guam

Prostrate shrub with trailing stems up to 3 m long. Leaves simple, alternate, blade somewhat fleshy, broadly ovate, sometimes lobed, 1.5–7 cm long on a petiole of similar length, margins toothed, surfaces velvety pubescent. Flowers several, in short, congested, terminal or leaf-opposed cymes. Calyx of 5 linear sepals 7–12 mm long, yellow inside. Corolla of 5 oblanceolate, yellow petals 8–11 mm long. Stamens many, yellow. Ovary superior. Fruit a subglobose bur 8–16 mm wide, covered with hooked spines.

Triumfetta procumbens is distributed from Micronesia and Malaysia (or perhaps as far east as the Seychelles) to eastern Polynesia, and is found on most of the high and low archipelagoes of Polynesia (except the Marquesas and Hawai'i) and Micronesia. It grows on sandy beaches, and is occasionally somewhat weedy in sandy coconut plantations, but rarely occurs very far inland.

In the Tuamotus, the bark fibers were used to make fishing lines, cordage, and ornamental skirts worn in dancing. The stems were soaked in seawater overnight and stripped of their fibrous bark. In Tokelau, the bark is sometimes used as a shampoo, the flowers are strung into leis, and the leaves and bark are employed in native medicines. The plant is hardly known elsewhere in Polynesia.

WOLLASTONIA BIFLORA
Asteraceae (Sunflower family)

English name: beach sunflower
Vernacular names: *titi tai*--Cooks (Atiu); *mata kula* – Niue; *ate* – Tonga; *ateate* – Samoa; *kovekove* – Fiji; *masiksik* – Guam

Large, erect to sprawling herb or subshrub up to 1.5 m in height, slightly woody at base, with angled, pubescent stems. Leaves simple, opposite, blade ovate, 6 – 15 cm or more long, appearing palmately 3-veined from the base, surfaces slightly hairy, margins toothed. Flowers in composite heads 8 – 15 mm wide, arranged in loose terminal panicles. Ray florets 6 – 10, yellow, 6 – 9 mm long; disc florets tubular, 20 – 30, yellow, 4 – 6 mm long. Fruit an obconical achene 2.5 – 4 mm long. SYNONYMS: *Wedelia strigulosa*, *Wedelia biflora*, *Wedelia canescens*.

Wollastonia biflora is distributed from tropical Asia to eastern Polynesia, and occurs on most of the high islands of Polynesia as far east as Rapa, and in the Marianas and Carolines of Micronesia. It grows in littoral thickets on rocky shores, sometimes being the dominant species in these habitats, but also becoming weedy in coastal coconut plantations.

This plant is commonly used for native medicines in western Polynesia; in Tonga, juice from the leaves is applied to cuts and wounds, a use also reported from Micronesia. In Samoa, an infusion of the leaves is commonly taken for urinary tract infections.

VINES

ABRUS PRECATORIUS
Fabaceae (Pea family)

English name: rosary pea
Vernacular names: *pūkiawe lei* – Hawai'i; *pitipiti'o* – Societies, Cooks; *pomea mata'ila* – Niue; *matamoso* – Samoa; *matamoho, moho* – Tonga; *leredamu* – Fiji; *kulales halom tano* – Guam

Thin-stemmed woody climber. Leaves alternate, pinnately compound, leaflets 16–34, opposite, blades subsessile, oblong, mostly 10–22 mm long, obtuse to notched at the tip, surfaces glabrous. Flowers many, in raceme-like panicles of spikes 4–12 cm long. Calyx cup-shaped, 3–4 mm long, shallowly 5-toothed. Corolla papilionaceous, lavender, 7–12 mm long. Stamens 9. Ovary superior. Fruit a flattened oblong pod 2–5 cm long with a short curved beak, enclosing several round seeds that are scarlet with a black area.

Abrus precatorius is distributed from tropical Africa to the Marquesas, and has been introduced and naturalized in America and Hawai'i. It is native or possibly an ancient introduction to most of the high islands of Polynesia, and to the Marianas and Carolines in Micronesia. The vine grows over littoral shrubs and into trees of littoral, coastal, and lowland forests.

The showy, red and black seeds of *Abrus* are useful for making seed leis, other types of decoration, and seed rattles, but they are poisonous if eaten. According to some sources, the plant was introduced into the Marquesas by Catholic missionaries for use in making rosaries.

CANAVALIA CATHARTICA
Fabaceae (Pea family)

Vernacular names: *maunaloa* – Hawai'i; *tutui faraoa?* – Societies; *kaka poti* – Cooks; *drautolu?* – Fiji; *lodosong tasi* – Guam

Prostrate vine to woody liana. Leaves alternate, trifoliate on a rachis 3 – 15 cm long, leaflet blades elliptic to ovate, 7 – 15 cm long, tip acute, surfaces glabrous. Flowers several, in axillary racemes 8 – 30 cm or more long. Calyx bilabiate, 10 – 15 mm long. Corolla papilionaceous, pink to magenta, 22 – 35 mm long. Stamens 10, diadelphous. Ovary superior. Fruit a flattened woody pod 7 – 12 cm long and 3 – 4 cm wide, each valve with a ridge running near and parallel to the edge, seeds bean-like, 10 – 16 mm long, hilum 2/3 or more as long. SYNONYMS: *Canavalia microcarpa, Canavalia turgida, Canavalia ensiformis* of some authors.

Canavalia cathartica is distributed from East Africa to Hawai'i (where it is a modern introduction), and is found in all the volcanic archipelagoes of Micronesia and Polynesia except the Marquesas. It grows on high islands, either prostrate near the shore or climbing as a liana in littoral forest, and is often weedy in inland plantations and disturbed forests.

No significant uses have been reported for this vine in Polynesia, and it is often not distinguished from other vines or lianas. A similar species endemic to Ra'iatea (Society Islands), *Canavalia raiateensis*, has shorter pods and much larger seeds.

CANAVALIA ROSEA
Fabaceae (Pea family)

Vernacular names: *feseka*—Niue; *fue fai va'a*—Samoa; *drautolu*—Fiji; *akankang tasi*—Guam

Prostrate or weakly climbing vine. Leaves alternate, trifoliate on a rachis 6–22 cm long, leaflet blades elliptic to nearly round, 5–15 cm long, tip mostly round, surfaces glabrous. Flowers several, in axillary racemes 12–20 cm or more long. Calyx bilabiate, 6–10 mm long. Corolla papilionaceous, pink to purple, 15–25 mm long. Stamens 10, diadelphous. Ovary superior. Fruit a narrow, flattened, woody pod 10–15 cm long, 2–3 cm wide, each valve with a ridge near and parallel to the edge, seeds bean-like, 10–15 mm long, hilum about one half as long. SYNONYMS: *Canavalia maritima, Canavalia obtusifolia.*

Canavalia rosea is pantropical in distribution and widespread on the high islands of Polynesia (Marquesas, Niue, Samoa, Tonga) and Micronesia (Marianas, Carolines, Marshalls). It mostly grows prostrate on sandy beaches and rocky shores, but sometimes climbs into low vegetation. It can be distinguished from the similar *Canavalia cathartica* by its more prostrate habit, rounded rather than pointed leaf tips, and longer, narrower pods containing seeds bearing a hilum extending only about half their length.

Like several other species of littoral vines, it is often not distinguished by Polynesians, who typically call it by the general name for vines. No significant uses have been reported for it in the area.

CANAVALIA SERICEA
Fabaceae (Pea family)

English name: silky jackbean
Vernacular names: *pōhue*—Hawai'i; *feseka sea*—Niue; *fue veli*—Tonga

Weakly climbing, pubescent vine. Leaves alternate, trifoliate, rachis 5—18 cm long, leaflet blades broadly elliptic to orbicular, 5—10 cm long, silky pubescent on the lower and sometimes on the upper surface. Flowers several, in axillary racemes up to 10 cm long. Calyx bilabiate, 12—18 mm long. Corolla papilionaceous, pink, 3—4.5 cm long. Stamens 10, diadelphous. Ovary superior. Fruit a flattened woody pod 10—15 cm long, up to 3 cm wide, each valve with a ridge near and parallel to the edge, seeds bean-like, 10—14 mm long, hilum about half as long.

Canavalia sericea is distributed from New Caledonia and Micronesia (Carolines, Marshalls) to Hawai'i (where it was introduced), and is found on all the high archipelagoes of Polynesia except the Marquesas. It is now very rare or perhaps extinct in the Society Islands, where it has not been reported this century. The vine grows mostly on rocky or sandy shores of high islands, lying prostrate or climbing over low vegetation near the shore.

Like several other vines, this species is not distinguished by the inhabitants of most of the islands where it grows, and is often referred to by the general name for vines. No significant uses have been reported for it in Polynesia. It differs from the other two littoral species of *Canavalia* by its larger flowers and the dense silky pubescence of its leaves.

CASSYTHA FILIFORMIS
Cassythaceae (Cassytha family)

Vernacular names: *kauna'oa pehu* – Hawai'i; *taino'a* – Societies; *tainoka* – Cooks; *feteinoa* – Niue; *fetai* – Samoa; *fatai* – Tonga; *wa vere lagi, wa lutu mai lagi?* – Fiji; *agasi, mai'agas* – Guam

Leafless, herbaceous, parasitic vine often forming tangled mats, with green to orange stems attaching to host plants by means of haustoria (suckers). Leaves simple, alternate, reduced to tiny scales. Flowers several, in short spikes 1–4 cm long. Calyx of 3 ovate sepals about 1 mm long, with 3 tiny bracts below. Corolla of 3 white elliptic petals 2–3 mm long. Fertile stamens 9. Ovary superior. Fruit globose, 6–9 mm long, green, enclosed within the fleshy floral tube, 1-seeded.

Cassytha filiformis is pantropical in distribution, and is found on all the major archipelagoes of Micronesia and Polynesia except the Marquesas. It usually grows as a parasite on littoral herbs and shrubs, and is rarely found very far from the shore, except in Hawai'i, where it sometimes covers trees in lowland and foothill forests.

On atolls, where it is common, *Cassytha* is sometimes woven into head leis, and the fruit is sometimes eaten by children. In the Tuamotus, the dried stems were placed under sleeping mats as cushioning. In Tokelau, the crushed stems are sometimes used as a shampoo or hair conditioner. An infusion of the stems is used to treat hemorrhoids in the Society Islands, and for treating convulsions of infants in the Cook Islands.

CUSCUTA SANDWICHIANA
Cuscutaceae (Dodder family)

English name: Hawaiian dodder
Vernacular names: *kauna'oa, kauna'oa kahakai* – Hawai'i

Trailing parasitic vine with slender yellow to yellow-orange stems less than 1 mm in diameter attaching to a host plant by means of haustoria (suckers). Leaves appearing absent, but reduced to tiny scales. Flowers several, in compact cymose clusters. Calyx 3–6 mm long, deeply divided into 4 or 5 triangular lobes nearly enclosing the corolla. Corolla campanulate, white, 2–5 mm long, divided into 4 or 5 ovate lobes. Stamens 5. Ovary superior. Fruit a subglobose capsule 3–4 mm in diameter with the dry corolla surrounding it, containing 1–4 subglobose seeds about 2 mm in diameter.

Cuscuta sandwichiana is endemic to Hawai'i, where it is reported from all the main islands except Kaho'olawe. It grows in tangled mats as a parasite on littoral plants in sunny coastal habitats and sometimes occurs up to 300 m elevation. A related species, *Cuscuta campestris*, is a weed reported in Hawai'i, Samoa, Niue, and the Cook Islands, where it may parasitize crop plants, especially legumes.

No significant uses have been reported for *Cuscuta sandwichiana*, but because it is a parasite, it may have an impact on native littoral vegetation in Hawai'i.

DALBERGIA CANDENATENSIS
Fabaceae (Pea family)

Vernacular name: *denimana*? – Fiji

Liana or sprawling shrub. Leaves alternate, odd-pinnately compound, rachis 3–6 cm long, leaflets alternate, 3–7, blades obovate, 1–4 cm long, tip rounded to slightly notched, surfaces mostly glabrous. Flowers several, in short, axillary racemes or narrow panicles 1–4 cm long. Calyx campanulate, 2.5–3.5 mm long, shallowly divided into 5 unequal lobes. Corolla papilionaceous, white, 5–8 mm long. Stamens 10, diadelphous. Ovary superior. Fruit a curved, oblong pod 2.5–3.5 cm long, mostly 1-seeded.

Dalbergia candenatensis is distributed from India eastward into Tonga and Micronesia (Marianas, Carolines). It grows in mangrove swamps and littoral shrubland, but is not very common, at least in Polynesia.

Other than occasional inclusion in native remedies, no significant uses have been reported for this vine. It is apparently not recognized in Tonga, and can be confused with *Abrus precatorius*, which has much smaller leaflets.

DERRIS TRIFOLIATA
Fabaceae (Pea family)

Vernacular names: *fue 'o'ona* – Samoa; *kavahaha* – Tonga; *duva* – Fiji; *bagen* – Guam

High climbing liana. Leaves alternate, odd-pinnately compound, rachis 8–15 cm long, leaflets mostly 5 or 7, alternate, blades ovate to oblanceolate, 3–9 cm long, surfaces glabrous. Flowers several, in dense, hanging racemes up to 10 cm or more long. Calyx cup-shaped, 2–3 mm long, unlobed. Corolla papilionaceous, pink to white, 6–9 mm long. Stamens 10, diadelphous. Ovary superior. Fruit a flattened, oblong pod 3–4.5 cm long, containing 1 or 2 seeds.

Derris trifoliata is distributed from East Africa to eastern Polynesia (Mangareva) and Micronesia (Marianas and Carolines); in Eastern Polynesia, it occurs only on Mangareva, Rapa, Rurutu, Ra'ivavae, and Mangaia, but is on most of the high islands of western Polynesia. It grows as a liana in littoral forest, and is only rarely found very far from the shore. The plants in French Polynesia differ from those to the west in having ovate to oblong instead of ovate to elliptic leaflets, and may be a distinct species that is found mostly in inland forests.

In former times the macerated plant was spread in the lagoon as a fish poison in Polynesia, but nowadays its use for this purpose has been taken over by other introduced, more chemically active species of *Derris*. The plant is poisonous to fish, but harmless to humans.

ENTADA PHASEOLOIDES
Fabaceae (Pea family)

English name: St. Thomas bean
Vernacular names: *kaka*—Cooks; *fue inu*—Samoa; *sipi, valai, pa'anga* (the seeds)—Tonga; *wa lai*—Fiji; *gayi, bayogon dankolo*—Guam

Large, high-climbing liana. Leaves alternate, bipinnately compound with a terminal, 2-branched tendril, rachis 2–9 cm long, pinnae opposite, leaflets opposite, 2–6, blades unequally-sided, ovate to elliptic, 4–10 cm long, surfaces glabrous. Flowers many, in dense, narrow, axillary spikes or panicles of spikes 15–25 cm long. Calyx cup-shaped, 1–2 mm long. Corolla divided to the base into 5 lanceolate petals 3–4 mm long, green. Stamens 10, white. Ovary superior. Fruit a large woody pod up to 1 m or more long, containing several hard, shiny brown, disc-shaped seeds up to 5 cm across. SYNONYM: *Entada scandens.*

Entada phaseoloides is distributed from tropical Asia to Micronesia (Marianas, Carolines, Marshalls) and to all of the high archipelagoes of Polynesia as far east as the Australs, with a small disjunct population in Hawai'i (Kaua'i). It grows as a high-climbing liana, sometimes with a trunk as thick as a tree, in littoral and lowland forest, often covering forest trees.

When cut, the thick but soft, woody stems exude a potable, watery sap that is drunk in times of need. The plant is also used in native medicines in Tonga. The most useful part of the plant, however, is the large seed that is strung into seed leis and other decorations, and was formerly used as a throwing piece in native games. The tough stems have been used for coarse cables and as jump ropes.

IPOMOEA IMPERATI
Convolvulaceae (Morning-glory family)

Vernacular name: *hunakai – Hawai'i*

Herbaceous vine with prostrate stems up to 4 m in length, rooting at the nodes. Leaves simple, alternate, blade variable, mostly ovate to oblong, 1 – 6 cm long, tip notched to bilobed, margins entire to irregularly lobed, surfaces glabrous. Flowers solitary in the leaf axils, on a peduncle 5 – 25 mm long. Calyx of 5 oblong, unequal sepals 7 – 15 mm long, on a pedicel 8 – 15 mm long. Corolla funnelform, white with a yellow throat, shallowly 5-lobed, 3.5 – 5 cm long. Stamens 5. Ovary superior. Fruit a globose to ovoid capsule 0.8 – 1.2 cm long, containing 1 or 2 seeds 7 – 9 mm long, covered with brown colored hairs. SYNONYM: *Ipomoea stolonifera*.

Ipomoea imperati is pantropical in distribution, but in the Pacific islands is restricted to Hawai'i, where it is reported from all the main islands except Hawai'i and Lana'i. It is uncommon on sandy beaches and coastal slopes down to just above the hightide mark.

No uses are reported for this species in Hawai'i, but the leaves and stems of *Ipomoea indica*, a related species of the lowlands, are used to treat wounds, sores, and broken bones.

IPOMOEA LITTORALIS
Convolvulaceae (Morning-glory family)

Vernacular names: *papati* – Societies, Cooks; *pipi* – Cooks (Rarotonga); *tefifi* – Niue; *palulu*? – Samoa; *lagon-tasi* – Guam

Herbaceous climbing vine. Leaves simple, alternate, blade cordate (rarely lobed), 2 – 10 cm long, surfaces glabrous. Flowers axillary, solitary, or in few-flowered cymes up to 7 cm long. Calyx of 5 unequal, oblong to suborbicular sepals 6 – 10 mm long, each with a short mucro. Corolla funnelform, lavender, shallowly 5-lobed, 3 – 4.5 cm long. Stamens 5. Ovary superior. Fruit a papery subglobose capsule 7 – 10 mm long, containing 4 seeds. SYNONYMS: *Ipomoea forsteri, Ipomoea denticulata, Ipomoea gracilis* of some authors.

Ipomoea littoralis is distributed from Madagascar to Hawai'i, and is recorded from all the high archipelagoes of Polynesia and all the high and low archipelagoes of Micronesia. It occurs from the shore to the lowlands, mostly in open sunny areas climbing over low vegetation, but is generally absent from atolls.

The only uses recorded for this somewhat weedy vine are medicinal. In the Society and Cook Islands, it is sometimes employed in native herbal remedies for some children's ailments known as *ira*.

IPOMOEA MACRANTHA
Convolvulaceae (Morning-glory family)

Vernacular names: *pō'ue* – northern Cooks; *fue sea* – Niue; *fue hina?* – Tonga; *wa ika?, wa mila?* – Fiji; *alaihai tasi* – Guam

Prostrate vine or somewhat woody liana with milky latex. Leaves simple, alternate, blade sometimes slightly fleshy, cordate to suborbicular (rarely lobed), 8 – 20 cm long, surfaces glabrous. Flowers solitary or sometimes 2 – 4 in short, axillary cymes. Calyx of 5 unequal, ovate sepals 15 – 25 mm long with a rounded tip. Corolla salverform, white, tube 6 – 10 cm long with an expanded, flattened, shallowly 10-lobed limb, opening at night. Stamens 5. Ovary superior. Fruit a papery, globose to ovoid capsule up to 2.5 cm long, containing 4 seeds. SYNONYMS: *Ipomoea tuba, Ipomoea violacea* of some authors, *Ipomoea grandiflora.*

Ipomoea macrantha is pantropical in distribution, and is found on nearly all of the high and low archipelagoes of Polynesia and Micronesia. It grows as a prostrate vine on the beaches of both high islands and atolls, or as a semi-woody liana in littoral forest. The recent flora of Hawaii (Wagner *et al.* 1990) refers to this species as *Ipomoea violacea*, but the more recent flora of Fiji (Smith 1991) retains the present name.

This is the most widespread of the native morning-glory species in the Pacific, but it is often not distinguished from other littoral vines and is usually called by the general name for vines. It has few reported uses, but is sometimes employed as a jump rope. In Tokelau, the leaves are used in native medicines for treating sores and rashes.

IPOMOEA PES-CAPRAE
Convolvulaceae (Morning-glory family)

English name: beach morning-glory
Vernacular names: *pōhuehue* — Hawai'i; *pōhue* — Societies; *pō'ue* — Cooks; *fue moa* — Samoa; *fue kula* — Tonga; *wa vulavula, lawere* — Fiji; *alalak tasi* — Guam

Trailing glabrous vine with purple stems, often rooting at the nodes. Leaves simple, alternate, blade fleshy, oblong to suborbicular, 3–10 cm long, notched at the tip, surfaces glabrous. Flowers axillary, solitary or in few-flowered cymes up to 15 cm long. Calyx of 5 unequal, ovate to elliptic sepals 8–13 mm long. Corolla funnelform, pink to rose-purple, 3–5 cm long, shallowly 10-lobed. Stamens 5. Ovary superior. Fruit an ovoid to subglobose capsule 12–17 mm long, containing 4 dark, ovoid, densely hairy seeds 6–10 mm long. SYNONYM: *Ipomoea brasiliensis.*

Ipomoea pes-caprae is pantropical in distribution, and is found on all the major high archipelagoes of Polynesia and all the low and high archipelagoes of Micronesia. It is one of the most abundant species on rocky and sandy beaches of high islands, sometimes forming almost pure stands, but is uncommon on atolls. On sandy beaches, its creeping stems extend almost down to the hightide mark.

Although this vine is widespread and common throughout Polynesia, few uses are reported for it. Often its name is the general one for vines (*pōhue, fue*), and it is often not distinguished from other littoral vines by the islanders.

JACQUEMONTIA OVALIFOLIA
Convolvulaceae (Morning-glory family)

English name: jacquemontia
Vernacular name: *pa'ū-ō-Hi'iaka* – Hawai'i

Prostrate vine with stems up to 3 m long, often rooting at the nodes. Leaves simple, alternate, blade thick, elliptic, suborbicular, 2–6 cm long, tip notched, surfaces densely pubescent, petiole 4–25 mm long. Flowers 1–few in an axillary cyme atop a long peduncle. Calyx of 5 unequal sepals 4–8 mm long, pedicel 5–10 mm long. Corolla broadly campanulate, pale blue to white, shallowly 5-lobed, 10–15 mm long. Stamens 5. Ovary superior. Fruit a subglobose capsule 4–6 mm in diameter enclosed within the enlarged sepals, containing 1–4 seeds. SYNONYM: *Jacquemontia sandwicensis.*

Jacquemontia ovalifolia is distributed from West Africa to the Pacific islands, but in Polynesia is found only in Hawai'i. The Hawaiian population, which is found on all the main islands, comprises an endemic subspecies (*sandwicensis*). It is common on rocky and sandy shores, and sometimes inland in sunny places up to 500 m in elevation.

No significant uses are reported for the plant, but since it is so common, it is often an important component of the littoral vegetation. Its Hawaiian name translates as "skirt of Hi'iaka." In Hawaiian mythology, Hi'iaka was the sister of the volcano goddess Pele, who gave the name to the plant when she returned from a fishing excursion to find the plant had grown over and shaded her baby sister.

MUCUNA GIGANTEA
Fabaceae (Pea family)

English name: sea bean
Vernacular names: *kā'e'e* – Hawai'i; *tutae pua'a* – Societies; *feseka uli* – Niue; *tupe* (the seed) – Samoa; *pa'anga 'ae kumā* (the seed), *valai* – Tonga; *wakore?* – Fiji; *bayogon dikike, gaogao dalalai* – Guam

Large woody climber. Leaves alternate, trifoliate, rachis mostly 8–16 cm long, leaflet blades broadly elliptic to ovate, 4–13 cm long, surfaces mostly glabrous. Flowers in axillary, hanging racemes up to 35 cm or more long. Calyx cup-shaped, bilabiate, 3–4 mm long. Corolla papilionaceous, yellow green, 3–4.5 cm long. Stamens 10, diadelphous. Ovary superior. Fruit a flattened, oblong pod 10–15 cm long, lacking transverse ridges on the flat surfaces, and containing 2–4 hard, brown, disc-shaped seeds 2–3 cm in diameter.

Mucuna gigantea is native from East Africa to Hawai'i, and is found in all the high archipelagoes of Polynesia and Micronesia (except the Carolines). It grows as a liana in littoral to lowland forest up to an elevation of about 200 m. Several other species of *Mucuna*, some of them more typically found inland, occur on many of the islands of Polynesia.

Lianas are often hard to distinguish from each other because their leaves are high in the canopy, and on many Polynesian islands *Mucuna* and other liana species (especially *Entada phaseoloides*) may be called by the same name. The smaller vines of *Mucuna* may be employed as jump ropes, and the hard brown seeds are used in seed leis.

VIGNA MARINA
Fabaceae (Pea family)

English name: beach pea
Vernacular names: *nanea* – Hawai'i; *pipi tatahi*? – Societies; *keketa, ka'eta, po'ue* – Cooks; *feseka tai* – Niue; *fue sina* – Samoa; *lautolu tahi* – Tonga; *drautolu* – Fiji; *akankang manulasa* – Guam

Prostrate or weakly climbing vine with longitudinally striate stems. Leaves alternate, trifoliate on a rachis 2 – 16 cm long, leaflet blades ovate to suborbicular, 4 – 12 cm long, surfaces nearly glabrous. Flowers many, in axillary racemes 5 – 25 cm long. Calyx cup-shaped, 3 – 5 mm long, with five triangular lobes. Corolla papilionaceous, yellow, 12 – 15 mm long. Stamens 10, diadelphous. Ovary superior. Fruit a black, narrowly cylindrical pod 4 – 8 cm long, containing many dark brown, bean-like seeds 5 – 7 mm long. SYNONYM: *Vigna lutea*.

Vigna marina is pantropical in distribution and is found on all the high archipelagoes of Polynesia and Micronesia. It is often dominant on sandy (and sometimes rocky) shores of high islands, growing prostrate or climbing over low vegetation; it occasionally occurs as a weed in open, inland habitats.

The plant is sometimes used as fodder for livestock, but its most important use is in native medicines. It is commonly employed in Tonga, Samoa, and the Cook Islands for treating "ghost sickness" (supernaturally induced ailments) and sometimes as a poultice for swellings and fractures.

GRASSES AND SEDGES

CENCHRUS CALYCULATUS
Poaceae (Grass family)

English name: Polynesian bur-grass
Vernacular names: *piripiri* – Societies; *hefa* – Tonga

Robust, erect annual grass. Culms up to 2 m in height, nodes prominent. Leaf blade narrowly lanceolate, 15–45 cm long, glabrous. Ligule tiny, membranous, with a fringe of hairs 0.5–1 mm long. Flowers in a spike-like raceme 6–27 cm long. Spikelets awnless, in clusters surrounded by a hairy, bur-like fruit 3–7 mm long, spines soft, 2–8 mm long, pointing upward. SYNONYM: *Cenchrus anomoplexis.*

Cenchrus calyculatus is distributed from New Caledonia to Pitcairn Island. In Polynesia, it is reported from the Society Islands, Cook Islands, Mangareva, Samoa, Tonga, and Niue, and from the Marshalls in Micronesia. It typically grows on rocky coasts and in coastal thickets, and occasionally inland in open sunny places. It is rare today throughout its Polynesian range, and may be extinct on most of the islands where it once occurred, perhaps being replaced by an introduced weedy species, *Cenchrus echinatus.*

A similar species, *Cenchrus agrimonioides*, is endemic to Hawai'i, where it occurs in the mountains; a littoral variety from the Leeward Hawaiian Islands is thought to be extinct. No significant uses have been reported for *Cenchrus calyculatus* grass, and because of its bur-like fruits, it has probably been more of a nuisance than anything else. It is probably dispersed by adhering to seabird feathers.

CYPERUS STOLONIFERUS
Cyperaceae (Sedge family)

Vernacular name: *pakopako* – Tonga; *mumuta?* – Samoa

Perennial sedge up to 50 cm in height. Culms erect, 3-angled, creeping by means of rhizomes, forming stout tubers 1–1.5 cm in diameter. Leaves mostly basal, linear, up to 30 cm or more long, 1–4 mm wide, margins entire or finely scabrid, surfaces glabrous. Flowers in several subglobose clusters of 3–10 sessile, flattened, brown spikelets 5–15 mm long and linear- lanceolate in outline. Florets 8–20 per spikelet. Stamens 3. Ovary superior, stigmas 3. Fruit an ovoid nut 1–1.5 mm long, dark brown to black, 1-seeded.

Cyperus stoloniferus is distributed from Madagascar eastward to western Polynesia, and is reported from Tonga and Samoa, but surprisingly not from Fiji. It may occur elsewhere, including Fiji, but since it is usually in a sterile state, it has often been misidentified. It grows on sandy beaches and in crevices on rocky coasts, sometimes becoming locally abundant.

In Tonga, the juice from the grated, fragrant tubers is used to scent coconut oil. A similar-looking species, the cosmopolitan weed *Cyperus rotundus*, is also used for this purpose.

FIMBRISTYLIS CYMOSA
Cyperaceae (Sedge family)

Vernacular names: none

Densely tufted sedge up to 30 cm in height. Culms erect, base covered with leaf sheaths and their remains. Leaves linear, stiff, mostly 2–12 cm long, 1–3 mm wide, glabrous. Flowers in heads 7–15 mm in diameter, or open corymbs 3–5 cm long. Spikelets brown, oblong, 3–6 mm long, many-flowered. Stamens 3. Ovary superior, style lobes 2 or 3. Fruit an obovoid, 3-angled achene 0.5–1 mm long. SYNONYMS: *Fimbristylis spathacea, Fimbristylis atollensis, Fimbristylis pycnocephala, Fimbristylis umbellato-capitata.*

Fimbristylis cymosa is distributed from Malaysia to Hawai'i and Micronesia, and eastward to tropical America; it occurs on most of the archipelagoes in this area except for the Marquesas. It is often the dominant and sometimes only species on rocky and sandy coasts, and only rarely occurs far from the shore.

Although this species is common, it is often not distinguished from other sedges and grasses, and is usually called by the general name for these. No significant uses are reported for it, other than its stem being employed to clean the ears.

ISCHAEMUM BYRONE
Poaceae (Grass family)

Vernacular names: none

Erect to ascending perennial grass. Culms up to 60 cm or more in height. Leaf blades linear, 3–25 cm long, glabrous. Ligule membranous, ovate to truncate, 1.5–3 mm long. Flowers in a pair of terminal, erect to ascending spikes 5–10 cm long, bearded with yellowish hairs and breaking apart into segments at maturity. Spikelets lanceolate in outline, paired, one sessile, one pedicellate, 4.5–7 mm long, 2-lobed at the attenuate tip, often hairy, bearing a bent awn 1.5–3 cm long. SYNONYMS: *Ischaemum lutescens, Ischaemum stokesii.*

Ischaemum byrone is endemic to eastern Polynesia, where it is reported only from the Austral Islands (Rapa and Rurutu), Cook Islands (Mangaia, Aitutaki, Miti'aro), and Hawai'i (Moloka'i, Maui, Hawai'i). It grows on rocky shores, and only occasionally away from the coast in sunny localities (Rapa), but nowhere is it reported to be common.

Like most other grasses, no local names or significant uses have been reported for this species. Because of its relative rarity, it is unlikely that islanders would even recognize it, and would probably refer to it by the general name for grasses. It is less common than other littoral species, and is related to another uncommon littoral grass of the same genus, *Ischaemum murinum,* found in western Polynesia (Niue, Tonga, Samoa).

LEPTURUS REPENS
Poaceae (Grass family)

Vernacular name: *lasaga* – Guam

Tufted perennial grass. Culms up to 30 cm or more in height, creeping, rooting at nodes. Leaf blade linear to linear-lanceolate, 3–15 cm long, glabrous, sometimes involute. Ligule membranous, tiny, with a fringe of hairs 0.4–0.8 mm long. Flowers in narrow, cylindrical spikes 5–15 cm long. Spikelets solitary, lanceolate in outline, attenutate-tipped, mostly 7–10 mm long, embedded in cavities and falling attached to a cylindrical section of the jointed spike. SYNONYMS: *Lepturus cinereus, Rottboellia repens.*

Lepturus repens is widely distributed from the Mascarene Islands in the Indian Ocean to eastern Polynesia, and occurs in nearly all the archipelagoes of Micronesia and Polynesia except the Marquesas and the main islands of Hawai'i. It is often the most abundant grass on rocky and sandy shores of atolls and high islands throughout the region, and only rarely grows very far inland.

Although this grass is often abundant, it is rarely given specific names in Polynesia, and is usually called by the general name for grasses. No significant uses are reported for it. Some botanists recognize several other species or varieties of *Lepturus*, but they are very hard to distinguish. One species, *Lepturus acutiglumis*, differs in having a more pronounced, long-attenuate awn on the spikelet, and is reported from Fiji and elsewhere.

MARISCUS JAVANICUS
Cyperaceae (Sedge family)

Vernacular names: *'ahu'awa* – Hawai'i; *mo'u ha'ari* – Societies; *mauku tatau tai* – Cooks; *selesele* – Samoa; *mahelehele* – Tonga; *davairadua* – Fiji

Coarse tufted sedge up to 1 m in height. Culms erect, 3-sided, with pith inside. Leaves many, basal, up to 80 cm long, 6 – 12 mm wide, often glaucous, edges finely scabrid. Flowers in spreading, compound corymbs 4 – 15 cm long. Spikeles mostly lanceolate in outline, 4 – 6 mm long, 4 – 6-flowered, brown. Stamens 3. Ovary superior, style 3-lobed. Fruit an ovate to obovate, 3-sided achene 1 – 2 mm long, brown. SYNONYMS: *Cyperus javanicus, Cyperus pennatus.*

Mariscus javanicus is distributed from tropical Africa to Hawai'i, and is found on most of the high and low archipelagoes of Polynesia and Micronesia. It is occasional on rocky and sandy coasts, in brackish swamps, in coastal taro fields, and even occasionally in the mountains up to an elevation of 300 m or more (Marquesas).

This sedge is often not distinguished from other sedge species, but in eastern Polynesia (Society and Cook Islands) the stem fibers are sometimes still used to squeeze grated coconut to extract the cream, and for filtering native medicines (Society Islands). The leaves and stems were also once employed in native medicines in the Society Islands and Hawai'i.

PASPALUM VAGINATUM
Poaceae (Grass family)

Vernacular name: *mohuku ano* – Tonga

Erect perennial grass. Culms up to 60 cm in height, spreading by means of creeping stolons, with leaf sheaths often covering the internodes. Leaf blade linear, 2.5–15 cm long, involute. Ligule less than 1 mm long, membranous, often with long white hairs behind it. Flowers in paired (rarely in 3s) terminal spike-like racemes 2.5–6 cm long. Spikelets solitary, ovate-lanceolate in outline, 2.5–5 mm long, awnless. SYNONYM: *Paspalum distichum* of some authors.

Paspalum vaginatum is pantropical in distribution, and is found in nearly all the Polynesian and Micronesian archipelagoes, perhaps as a modern, natural arrival. It is common on sandy and rocky shores, and is often the dominant species along estuaries and on the edges of mangrove swamps, but is rarely found far from the shore.

Like most of the other Polynesian grasses, it usually has no specific name and is called locally by the general term for grasses. No significant uses have been reported in Polynesia. It has often been identified as *Paspalum distichum*, but this is probably a distinct species, rare if at all present in Polynesia. *Paspalum vaginatum* is easily confused with *Sporobolus virginicus* because the two species dominate littoral habitats and are usually found in a sterile condition. Perhaps the best distinguishing character is the ligule, which is tiny (but with a fringe of hairs present) in *Sporobolus*, and distinct and membranous in *Paspalum*.

SPOROBOLUS VIRIGINICUS
Poaceae (Grass family)

English name: beach dropseed
Vernacular names: *'aki'aki* – Hawai'i; *hatopa* – Guam

Erect, wiry perennial grass. Culms 5–50 cm in height, spreading by means of scaly underground rhizomes. Leaf blade linear, usually involute, mostly 3–12 cm long, sheaths overlapping, mostly glabrous. Ligule tiny, with a fringe of hairs about 0.5 mm long. Flowers in dense, narrow, cylindrical, spike-like panicles 2–9 cm long. Spikelets lanceolate in outline, laterally compressed, 2–3 mm long, acute-tipped, glabrous, 1-flowered.

Sporobolus virginicus is widely distributed in the tropics and subtropics of the Old and New World, but is scattered in the Pacific islands, where it is reported from Hawai'i (all the main islands), Tonga, Fiji, and Micronesia (all the main archipelagoes). It has until recently been misidentified as *Spinifex littoreus* in Tonga. It grows on coastal sand dunes and sandy shores down to just above the hightide mark, and rarely inland at up to 100 m elevation.

No uses have been reported for this grass, but it serves a useful purpose in stabilizing sand dunes, and is hence an important part of the vegetation. It is usually found in a sterile condition, and when so, is very similar to *Paspalum vaginatum*, which is found in similar habitats, but more often on the edges of estuaries.

STENOTAPHRUM MICRANTHUM
Poaceae (Grass family)

Vernacular names: none

Tufted, erect to ascending, perennial grass. Culms up to 40 cm in height, creeping, rooting at the prominent nodes. Leaf blade lanceolate, mostly 2.5–10 cm long, glabrous. Ligule short, with a dense fringe of hairs up to 1 mm long. Flowers in slender, corky, cylindrical spikes 2.5–20 cm long. Spikelets oblong to lanceolate in outline, 1–2.5 mm long, 2–4 on short branches, embedded in cavities on one side of the spike, which remains whole at maturity. SYNONYM: *Stenotaphrum subulatum.*

Stenotaphrum micranthum is distributed from the Mascarene Islands in the Indian Ocean to eastern Polynesia, and is widespread in Polynesia (the Society Islands, Austral Islands, Cook Islands, Samoa, and Tonga) and Micronesia (Marianas, Carolines, Marshalls). It grows on sandy shores and sometimes in open littoral forest, but never far from the shore. Over much of its range it is uncommon and easily mistaken for other littoral grass species, particularly *Lepturus repens*, a dominant species in the same habitats; *Lepturus* differs in having a jointed rachis breaking into segments at maturity.

Stenotaphrum has no reported native names in Polynesia, and is called locally by the general name for grasses. No significant uses have been reported for it.

THUARAEA INVOLUTA
Poaceae (Grass family)

Vernacular names: *kefukefu* – Tonga; *lasaga*? – Guam

Prostrate perennial grass. Culms creeping, branching, up to 1 m or more in length, rooting at nodes. Leaves in 2 rows on the stem, blade lanceolate to linear-lanceolate, 2 – 8 cm long, mostly glabrous. Ligule short, with a fringe of hairs about 1 mm long. Flowers in 1-sided, terminal, few-flowered spikes enclosed by a folded, lanceolate sheath 1.2 – 2.5 cm long. Spikelets ovate, sessile, 3 – 5 mm long, enclosed at maturity by the rolled-up sheath which forms a watertight case 6 – 10 mm long. SYNONYM: *Ischaemum involutum.*

Thuarea involuta is distributed from Malaysia eastward to eastern Polynesia (Henderson Island), and is found in nearly all the archipelagoes of Polynesia (except the Marquesas and Hawai'i) and Micronesia. It grows on sandy beaches and coral rubble near the shore, more frequently on high islands than on atolls, and is often a dominant species where it occurs.

This species is usually not distinguished by Polynesians from the other littoral grass species. In Tonga, however, a crude lei of this grass is worn by fishermen who are presenting fish to a chief.

GLOSSARY OF BOTANICAL TERMS

Achene— A small, dry, non-splitting, 1-seeded fruit with the seed fused to the ovary wall, as in the grass family.

Acute— Tapering to a sharp, but not drawn-out, point. Compare attenuate.

Alternate— Said of leaves arranged one per node. Compare opposite.

Annual— Having a life span of a single year or season. Compare perennial.

Anther— A pollen sack of stamen. See filament.

Anthocarp— A specialized type of achene.

Appressed— Lying flat against a surface, as appressed hairs.

Attenuate— Tapering gradually to form a long, straight-sided tip. Compare acute.

Awn— A slender, bristle-like appendage usually at the tip of a structure, especially on grass spikelets.

Axil— The upper angle between a leaf petiole and stem. Flowers situated in the axil are referred to as being axillary.

Berry— A fleshy, pulpy fruit containing two or more seeds, such as a grape, passionfruit, or guava.

Bilabiate— Two-lipped, said of a corolla or calyx with the parts fused into an upper and lower lip.

Bipinnately compound— Twice pinnate, the first divisions being further divided into leaflets.

Bract— A reduced or modified leaf subtending a flower or inflorescence. A secondary bract is sometimes called a bracteole.

Calyx— The outer, usually green whorl of the flower, enclosing the flower bud. It is composed of free or fused sepals.

Campanulate— Bell-shaped, said of a corolla or calyx.

Capitate— Head-shaped, or said of dense, globose inflorescences.

Capsule— A dry, splitting fruit with several cells, opening by sections called valves.

Circumscissile— Describing a fruit whose top splits off along a seam.

Composite— Composed of two types of florets, said of the flowers of the sunflower family, Asteraceae (Compositae). See disc and ray floret.

Compound— Said of leaves with the blade further divided into leaflets or pinnae. Compare simple.

Cordate— Heart-shaped, said of leaves.

Corolla – The inner, usually colored, whorl of a flower, composed of free or fused petals, or sometimes absent.
Coriaceous – Having a leathery texture, said of leaves.
Corymb – A flat-topped, short, broad inflorescence with the center flower the youngest.
Culm – The "stem" of a grass or sedge.
Cyme – A cluster of flowers with the oldest ones at the end or center. Compare panicle.

Decumbent – Prostrate, but with ascending stem tips.
Dehiscent – Splitting, said of fruits such as capsules.
Dichotomous – Dividing into two branches.
Dioecious – Condition of plants with unisexual flowers in which the male and female flowers are on separate plants. Compare monoecious.
Diadelphous – Bearing stamens in two bundles, especially in papilionaceous flowers that usually have a 9 plus 1 arrangement. Compare monadelphous.
Disc floret – A central, tubular flower of a "composite" inflorescence of the sunflower family (Asteraceae or Compositae). See ray floret.
Drupe – A fleshy fruit with a single seed enclosed in a hard shell, such as a mango or peach; "stone" fruit.

Elliptic – Shaped like an ellipse. An ellipsoid is a 3-dimensional figure shaped in outline like an ellipse, said of some fruits.
Entire – Having a continuous margin lacking teeth or lobes.
Epipetalous – Borne on the corolla or petals, said of stamens.
Exserted – Sticking out, said of stamens when they protrude from the corolla.

Fascicle – A condensed cluster, said of leaves or flowers.
Filament – The stalk of a stamen, bearing the anther.
Floret – A small flower of members of the sunflower family (Asteraceae or Compositae). See disc and ray florets.
Funnelform – Funnel-shaped, said of corollas of flowers.

Glabrous – Said of a surface lacking pubescence; hairless.
Glaucous – Covered with a bloom, or a white substance that rubs off.
Globose – Spherical in shape.

Hilum – The scar of the attachment point of a seed.

Inferior— Said of an ovary or fruit which has the sepals on top, i.e., the ovary is inferior to the attachment of the sepals. Compare superior.
Inflorescence— A flower cluster or the arrangement of flowers on a plant.
Interpetiolar— Situated on a stem between two petioles, said of stipules.
Involucre— A whorl of leaves or bracts close to the base of a flower cluster.
Involute— Rolled inward or toward the upper surface, as of some grass leaves. Compare revolute.

Lanceolate— Lance-shaped in outline, several times longer than wide, with the widest portion towards the base of the leaf. Compare oblanceolate.
Leaflet— A division of a compound leaf.
Ligule— A projection at the top of the leaf sheath in many grasses.
Limb— Expanded terminal portion of some corollas.
Linear— Long and narrow, with the sides almost parallel.

Monadelphous— Said of stamens united by their filaments into a single bundle. Compare diadelphous.
Monoecious— Condition of a plant with unisexual flowers when male and female flowers are on the same plant. Compare dioecious.
Mucro— A sharp, tooth-like tip of some leaves, bracts, petals, or other parts.

Node— Point of attachment of a leaf on a stem.
Nutlets— A small, 1-seeded, non-splitting lobe of a divided fruit.

Obconical— Shaped like a cone, with the broadest end at the tip.
Oblanceolate— Lanceolate, but with the widest part towards the tip. Compare lanceolate.
Oblique— Unequally-sided, as in the bases of some leaves.
Oblong— Longer than broad, with the sides nearly parallel to each other.
Obovate— Ovate, but with widest part towards the tip. Compare ovate.
Obtuse— Blunt, rounded.
Opposite— Referring to leaves borne in pairs at the node. Compare alternate.

Orbicular— Round in outline. Compare suborbicular.

Ovary— The female part of the flower, containing the embryos or immature seeds.

Ovate— Oval in outline, with widest part towards the base. Compare obovate.

Ovoid— Said of fruits, etc., which are oval in outline.

Palmate— Lobed or divided in a hand-like fashion, usually in reference to leaf blades or their veins. Compare pinnate.

Panicle— A compound inflorescence with a main axis and racemose branches, with the youngest flowers towards the tip. Compare cyme.

Papilionaceous— Butterfly-like, said of a sweetpea type of flower of the legume family (Fabaceae or Leguminosae).

Pedicel— The stalk of a flower.

Peduncle— The stalk of a flower cluster.

Petaloid— Petal-like in texture or color, said of the calyx of some flowers.

Peltate— Referring to a leaf that has the petiole joined to the blade away from or inside the margin.

Perennial— Living more than one season or year. Compare annual.

Perianth— The collective term for corolla and calyx.

Petal— A division of a corolla.

Petiole— The stalk of a leaf.

Phalange— One of the numerous woody sections of a pandanus fruit, containing the seeds. See syncarp.

Pinna— The first division of a compound leaf; the pinna is further divided into leaflets. Pinnae is the plural.

Pinnate— Divided in feather-like fashion. Compare palmate.

Pubescent— Hairy, covered with soft hairs.

Raceme— Simple, elongated inflorescence with stalked flowers on a single main axis (rachis), the youngest ones at the top.

Rachis— The axis of a compound leaf or inflorescence.

Ray floret— A strap-shaped flower of a "composite" inflorescence, usually in members of the sunflower family (Asteraceae or Compositae). Compare disc floret.

Revolute— Having the margins rolled towards the lower surface, said of leaves. Compare involute.

Rhizome— An underground stem, usually with nodes and buds.

Salverform— Said of a corolla with a slender tube and abruptly expanded limb.
Scabrid— Having a rough or finely serrate edge or surface.
Scapose— Said of an inflorescence having a long stalk (scape).
Schizocarp— A dry fruit splitting apart at maturity into 1-seeded segments.
Scorpeoid cyme— A cyme coiled like the tail of a scorpion.
Sepal— A division of the calyx.
Serrate— Having a saw-toothed margin, as of some leaves.
Sessile— Lacking a stalk, said of leaves, flowers, etc.
Simple— Said of a leaf or other structure that is not divided into parts. Compare compound.
Spathulate— Spoon-shaped.
Spike— An unbranched inflorescence bearing sessile flowers, the youngest ones at the tip.
Spikelet— A grass or sedge inflorescence consisting of flowers and membranous scales.
Stamen— The male part of the flower, consisting of an anther and a filament.
Staminode— A sterile stamen, lacking the anther.
Stellate— Star-shaped, said of some hairs.
Stigma— The sticky receptive tip of an ovary, with or without a style.
Stipe— The stalk of an ovary, in a few kinds of flowers.
Stipules— Paired basal appendages present on the petioles of some plants.
Stolon— A horizontal stem rooting at the nodes or producing a new plant at its tip.
Striate— Finely grooved.
Style— The stalk between the ovary and stigma.
Suborbicular— Nearly round in outline. Compare orbicular.
Subglobose— Nearly spherical in shape.
Subsessile— Nearly stalkless or sessile.
Subumbellate— Nearly umbellate. See umbel.
Superior— Said of an ovary or fruit with the sepals on the bottom. Compare inferior.
Sympetalous— Having the petals fused together to form the corolla.
Syncarp— The large, compound, woody fruit of pandanus, composed of numerous phalanges. See phalange.
Synsepalous— Having the sepals fused together to form the calyx.

Tepal— The collective term for sepals and petals when they are undifferentiated in a flower.
Terminal— Situated at the end of a branch or rachis.
Throat— The part of a sympetalous corolla where the limb joins the tube.
Tomentose— Densely woolly with tangled, matted hairs.
Toothed— Bearing teeth or indentations along the margin, said of leaf margins.
Trifoliate— Bearing leaves divided into three leaflets.
Truncate— Appearing cut-off or squared at the end, as of a leaf tip.
Tube— Basal, cylindrical portion of a corolla having fused petals.
Tuber— An underground, swollen, root-like stem of some plants.

Umbel— A flat- or round-topped inflorescence with the stalks of the flowers all arising from one point, the oldest flowers at the center.
Unisexual— Said of flowers lacking either stamens or an ovary.
Utricle— A bladder-like, 1-seeded fruit of some plants.

Valve— A section or piece into which a capsular fruit splits.

Whorled— Said of leaf arrangements having more than two leaves per node.
Winged— Having ridges, said of some stems with longitudinal ridges running down the stem or petiole.

SELECTED BIBLIOGRAPHY

Brown, F. B. H. 1931, 1935. Flora of southeastern Polynesia. B. P. Bishop Museum Bull. 84: 1–194 (I. Monocotyledons); 130: 1–386 (II. Dicotyledons).

Christophersen, E. 1935, 1938. Flowering plants of Samoa. B. P. Bishop Museum Bull. 128: 1–221; II. 154: 1–77.

Drake del Castillo, E. 1893. Flore de la Polynésie française. G. Masson, Paris. 351 pp.

Fosberg, F. R., M.-H. Sachet, and R. Oliver. 1979. A geographical checklist of the Micronesian Dicotyledonae. Micronesica 15: 1–295.

Fosberg, F. R., M.-H. Sachet, and R. Oliver. 1982. A geographical checklist of the Micronesian Monocotyledonae. Micronesica 18 (1): 23–82.

Merlin, M. D. 1977. Hawaiian coastal plants and scenic shorelines. Oriental Publishing Company. 68 pp.

Morat, P. and J.-M. Veillon. 1985. Contribution à la connaissance de la végétation et de la flore de Wallis et Futuna. Adansonia 3: 259–329.

Petard, P. 1986. Plantes utiles de Polynésie: ra'au Tahiti. Haere Po no Tahiti, Pape'ete, Tahiti. 354 pp.

Raulerson, L. and A. Rinehart. 1991. Trees and shrubs of the Northern Mariana Islands. Coastal Resource Management, Saipan. 120 pp.

Smith, A. C. 1979–1991. Flora Vitiensis nova; a new flora of Fiji. National Tropical Botanical Garden. 5 vols.

St. John, H. 1973. List and summary of the flowering plants in the Hawaiian Islands. Pacific Tropical Botanical Garden. 519 pp.

St. John, H. and A. C. Smith. 1971. The vascular plants of the Horne and Wallis Islands. Pacific Science 25 (3): 313–348.

Stone, B. 1970. Flora of Guam. Micronesica 6: 1–659.

Sykes, W. R. 1970. Contributions to the flora of Niue. New Zealand Dept. Sci. & Indust. Res. Bull. 200: 1–321.

Wagner, W. L., D. Herbst, and S. H. Sohmer. 1990. Manual of the flowering plants of Hawai'i. University of Hawaii Press & Bishop Museum Press, Honolulu. 2 vols.

Whistler, W. A. 1980. Coastal flowers of the tropical Pacific. Pacific Tropical Botanical Garden. 83 pp.

Whistler, W. A. 1984. Annotated list of Samoan plant names. Economic Botany 38 (4): 464–489.

Whistler, W. A. 1988. Ethnobotany of Tokelau: the plants, their Tokelau names, and their uses. Economic Botany 42 (2): 155–176.

Whistler, W. A. 1990. Ethnobotany of the Cook Islands: the plants, their Maori names, and their uses. Allertonia 5 (4): 347–424.

Whistler, W. A. 1991. Ethnobotany of Tonga: the plants, their Tongan names, and their uses. Bishop Museum Series in Botany 2: 1–155.

Whistler, W. A. 1992. Polynesian herbal medicine. National Tropical Botanical Garden, Lawai, Kaua'i, Hawai'i. In Press.

Wilder, G. P. 1931. Flora of Rarotonga. B. P. Bishop Museum Bull. 86: 1–113.

Yuncker, T. G. 1959. Plants of Tonga. B. P. Bishop Museum Bull. 220: 1–343.

INDEX TO SCIENTIFIC NAMES

Names in *italics* are synonyms. Those with an asterisk are littoral species mentioned but not featured in the text.

Abrus precatorius 113
Acacia laurifolia 19
Acacia simplex 19
Acacia simplicifolia 19
Achyranthes aspera* 77
Achyranthes atollensis* 76
Achyranthes mutica* 76
Achyranthes splendens 76
Achyranthes velutina 77
Argusia argentea 45
Atriplex semibaccata 78
Atriplex suberecta* 78

Bacopa monnieri 79
Barringtonia asiatica 20
Barringtonia speciosa 20
Batis maritima 48
Bikkia tetrandra 49
Boerhavia acutifolia 80
Boerhavia albiflora* 80
Boerhavia coccinea* 81
Boerhavia diffusa 80, 81
Boerhavia diffusa
var. *tetrandra* 82
Boerhavia herbstii* 80
Boerhavia glabrata 80
Boerhavia repens 81
Boerhavia tetrandra 82
Bramia monniera 79
Bruguiera conjugata 21
Bruguiera gymnorrhiza 21
Bruguiera rheedii 21

Caesalpinia bonduc 50
Caesalpinia bonducella 50
Caesalpinia crista 50
Caesalpinia major* 50
Calophyllum inophyllum 22
Canavalia cathartica 114
Canavalia ensiformis 114
Canavalia maritima 115
Canavalia microcarpa 114
Canavalia obtusifolia 115
Canavalia raiateensis* 114
Canavalia rosea 115
Canavalia sericea 116
Canavalia turgida 114
Capparis cordifolia 51
Capparis mariana 51
Capparis sandwichiana* 51
Capparis spinosa
var. *mariana* 51
Carapa moluccensis 47
Carapa obovata 46
Cassytha filiformis 117
Casuarina equisetifolia 23
Casuarina litorea 23
Cenchrus agrimonioides* 129
Cenchrus anomoplexis 129
Cenchrus calyculatus 129
Cenchrus echinatus* 129
Cerbera lactaria 24
Cerbera manghas 24
Cerbera manghas 25
Cerbera odollam 25
Chamaesyce atoto 83
Chamaesyce chamissonis 83
Chamaesyce chamissonis* 83

Chamaesyce degeneri 84
Chamaesyce skottsbergii* 84
Chenopodium oahuense 52
Chenopodium sandwicheum 52
Clerodendrum inerme 53
Cocos nucifera 26
Colubrina asiatica 54
Corchorus torresianus 55
Cordia subcordata 27
Cressa cretica 85
Cressa insularis 85
Cressa truxillensis 85
Cuscuta sandwichiana 118
Cyperus javanicus 134
Cyperus pennatus 134
Cyperus stoloniferus 130

Dalbergia candenatensis 119
Dendrolobium umbellatum 56
Derris trifoliata 120
Desmodium umbellatum 56

Entada phaseoloides 121
Entada scandens 121
Erythrina indica 29
Erythrina fusca 28
Erythrina variegata 29
Eugenia rariflora 57
Eugenia reinwardtiana 57
Euphorbia atoto 83
Euphorbia chamissonis 83
Euphorbia degeneri 84
Euphorbia ramosissimum 83
Euphorbia tahitensis 83
Excoecaria agallocha 30

Ficus scabra 58
Fimbristylis atollensis 131
Fimbristylis cymosa 131
Fimbristylis pycnocephala 131
Fimbristylis spathacea 131
Fimbristylis umbellato-capitata 131

Gnaphalium hawaiiense 86
Gnaphalium sandwicensium 86
Gossypium hirsutum 59
Gossypium religiosum 59
Gossypium tomentosum* 59
Guettarda speciosa 31

Haloragis prostrata 87
Hedyotis biflora 88
Hedyotis foetida 89
Hedyotis paniculata 88
Hedyotis romanzoffiensis 90
Heritiera littoralis 32
Hernandia nymphaeifolia 33
Hernandia ovigera 33
Hernandia peltata 33
Hernandia sonora 33
Hibiscus tiliaceus 34
Heliotropium anomalum 91
Heliotropium curassavicum 92

Ipomoea brasiliensis 125
Ipomoea denticulata 123
Ipomoea forsteri 123
Ipomoea gracilis 123
Ipomoea grandiflora 124
Ipomoea imperati 121
Ipomoea littoralis 123
Ipomoea macrantha 124
Ipomoea pes-caprae 125
Ipomoea stolonifera 122
Ipomoea tuba 124
Ipomoea violacea 124
Ischaemum byrone 132
Ischaemum involutum 138

Ischaemum lutescens 132
Ischaemum murinum* 132
Ischaemum stokesii 132

Jacquemontia ovalifolia 126
Jacquemontia sandwicensis 126
Jossinia reinwardtiana 57

Lepidium bidentatum 93
Lepidium bidentoides 93
Lepidium o-waihiense 93
Lepidium piscidium 93
Lepturus acutiglumis* 133
Lepturus cinereus 133
Lepturus repens 133
Leucaena forsteri 41
Leucaena insularum 41
Lipochaeta integrifolia 94
Lumnitzera littorea 35
Lycium carolinense var. *sandwicense* 60
Lycium sandwicense 60
Lysimachia mauritiana 95
Lysimachia rapensis* 95

Mariscus javanicus 134
Messerschmidia argentea 45
Mucuna gigantea 127
Myoporum boninense* 61
Myoporum rapense* 61
Myoporum sandwicense 61
Myoporum stokesii* 61
Myoporum wilderi 61

Nama sandwicensis 96
Neisosperma oppositifolium 36
Nesogenes euphrasioides 97
Nicotiana fatuhivensis* 98
Nicotiana fragrans 98

Ochrosia oppositifolia 36
Ochrosia parviflora 36
Oldenlandia biflora 88
Oldenlandia paniculata 88

Pandanus odoratissimus 37
Pandanus tectorius 37
Pariti tiliaceus 34
Paspalum distichum 135
Paspalum vaginatum 135
Pemphis acidula 62
Phyllanthus marianus* 63
Phyllanthus societatis 63
Pisonia grandis 38
Portulaca australis 100
Portulaca hawaiiensis 101
Portulaca johnii 99
Portulaca lutea 99
Portulaca molokiniensis* 99
Portulaca oleracea* 99
Portulaca pilosa 100, 101
Portulaca samoensis 100
Portulaca villosa 101
Premna gaudichaudii 64
Premna integrifolia 64
Premna obtusifolia 64
Premna serratifolia 64
Premna taitensis 64

Rhizophora apiculata* 40
Rhizophora mangle 39
Rhizophora mucronata 40
Rhizophora samoensis 39
Rhizophora stylosa 40
Rottboellia repens 133

Scaevola coriacea 65
Scaevola frutescens 66
Scaevola koenigii 66
Scaevola sericea 66

Scaevola taccada 66
Schiedea globosa 102
Schleinitzia insularum 41
Sesbania arborea 67
Sesbania atollensis* 67
Sesbania coccinea* 67
Sesbania hawaiiensis 67
Sesbania hobdyi 67
Sesbania molokaiensis 67
Sesbania tomentosa 67
Sesuvium portulacastrum 103
Sida cordifolia 68
Sida fallax 68
Sida meyeniana 68
Solanum amicorum 104
Solanum laysanense 105
Solanum nelsonii 105
Solanum viride* 104
Sophora tomentosa 69
Sporobolus virginicus 136
Stenotaphrum micranthum 137
Stenotaphrum subulatum 137
Suriana maritima 70

Tacca leontopetaloides 106
Tacca pinnatifida 106
Tephrosia mariana* 107
Tephrosia piscatoria 107
Tephrosia purpurea 107
Terminalia catappa 42
Terminalia glabrata* 42
Terminalia litoralis* 43
Terminalia microcarpa 43
Terminalia saffordii 43
Terminalia samoensis 43
Tetragonia expansa 108
Tetragonia tetragonioides 108
Tetramolopium rockii 109
Tetramolopium sylvae* 109
Thespesia populnea 44
Thuarea involuta 138
Timonius forsteri 71
Timonius polygamus 71
Tournefortia argentea 45
Tribulus cistoides 110
Triumfetta procumbens 111

Vigna lutea 128
Vigna marina 128
Vitex negundo var. *bicolor* 73
Vitex ovata 72
Vitex rotundifolia 72
Vitex trifolia 73
Vitex trifolia 72
Vitex trifolia var. *simplicifolia* 72

Wedelia biflora 112
Wedelia canescens 112
Wedelia strigulosa 112
Wollastonia biflora 112

Ximenia americana 74
Ximenia elliptica 74
Xylocarpus granatum 46
Xylocarpus moluccensis 47
Xylocarpus obovatus 46
Xylosma orbiculatum 75

INDEX TO VERNACULAR NAMES

A'abang 57
'Aerofai 77
Agasi 117
'Aheahea 52
Ahgao 64
Ahu'awa 134
'A'ie 62
'Aito 23
Akankang manulasa 128
Akankang tasi 115
Akataha 81
'Aki'aki 136
'Akoko 84
'Akulikuli 103
'Akulikuli kai 48
Alaihai tasi 124
Alalak tasi 125
Alena 80, 81
Aloalo 64
Aloalo tai 53
'Amae 44
'Anaoso 50
'Anapanapa 54
'Anaunau 93
'Ano 31
'Ara 37
'Atae 29
Ate 112
Ateate 112
'Ati 22
'Atoto 83
'Aturi 99
'Au 34
'Auhola 107
'Auhuhu 107
Autara'a 42, 43
'Avaro 64
'Avasa 107
'Aweoweo 52

Bagen 120
Bakau-aine 35
Balawa 37
Banalu 44
Bayogon 121
Bayogon dikike 127
Buabua 31
Buka 38
Butabuta 30

Dabi 46, 47
Dafao 82
Dalalai 127
Dankolo 121
Da'ok 22
Davairadua 134
Denimana 119
Dilo 22
Dogo 21, 39, 40
Drala dina 29
Drala kaka 28
Drala sala 73
Drautolu 114, 115, 128
Duva 120

'Ena'ena 86
Evu 45
Evuevu 33

Fa 37
Fago 36
Fala 37
Fara 37
Fao 36
Fatai 117
Fau 34
Feifai 41
Feseka 115
Feseka sea 116
Feseka tai 128
Feseka uli 127
Fetai 117
Feteinoa 117
Feta'anu 30
Fetanu 30
Fetau 22
Feta'u 22
Fotulona 33
Fou 34
Fihoa 54
Fisoa 54
Fue fai va'a 115
Fue hina 124
Fue inu 121
Fue kula 125
Fue moa 125
Fue 'o'ona 120
Fue sea 124
Fue sina 128
Fue veli 116
Fululupe 75
Futu 20

Gagu 23
Gaogao 29, 127

Gaosali 49
Gapgap 106
Gasoso 54
Gatae 29
Gatae palagi 28
Gayi 121
Gigia 62, 70

Ha'ari 26
Hala 37
Hangale 35
Hau 34
Hatopa 136
Hefa 129
Hinahina 91
Hinahina kahakai 96
Hola 107
Hora 107
Horahora 93
Hunakai 122
Hunek 45
Hutu 20

'Ihi 99, 101
'Ilima 68

Ka'e'e 127
Ka'eta 128
Kafo 37
Kaka 121
Kakalaioa 50
Kaka poti 114
Kamani 22
Kamani haole 42
Kamole 99
Katule 81
Kauariki 42
Kauna'oa 118
Kauna'oa kahakai 118
Kauna'oa pehu 117
Kautokiaveka 90
Kavahaha 120
Kavahuhu 107
Kaveutu 71
Kefukefu 138
Keketa 128
Kinakina 89
Kipukai 92
Kohuhu 107
Kopara 71
Koporoporo 90
Kou 27
Kovekove 112
Kulales halom tano 113
Kukuku 54
Kuru 70

Lagon-tasi 123
Lagunde 73
Lala 56
Lalanyok 47
Lala sea 73
Lalapa 28
Lala tahi 73
Lala'uta 56
Lasaga 133, 138
Lautolu tahi 128
Lawere 125
Legilegi 46, 47
Le'ile'i 47
Lekileki 46, 47
Leredamu 113
Leva 24, 25
Liki 57, 75
Lodigao 53
Lodosong tasi 114

Mahelehele 134
Mahoa'a 106
Mai'agas 117
Mamea 32
Mangle hembra 40
Mangle machu 21
Masi 58
Masiksik 112
Masiksik hembra 111
Masoa 106
Mata kula 112
Matamoho 113
Matamoso 113
Mata'ora 107
Mati 58
Mauku tatau tai 134
Maunaloa 114
Milo 44
Miro 44
Moemoe 82
Mohuku ano 135
Moli tahi 74
Moli tai 74
Motou 27
Mo'u ha'ari 134
Mulomulo 44
Mumuta 130

Naio 61
Namulega 73

Nanasu 66
Nanea 128
Nau 93
Naunau 93
Naupaka 65
Naupaka kahakai 66
Naupata 66
Nawanawa 27
Nehe 94
Nena 92
Nietkot 62, 70
Nigas 62
Nioi 57
Niyok 26
Niyoron 27
Niu 26
Ngahu 66
Ngahupa 66
Ngaio 61
Nganga 35
Ngangie 62, 70
Ngatae 29
Ngatae fisi 28
Ngate 29
Nga'u 66
Ngingie 62, 70
Nohu 110
Nokonoko 23
Nonak 33
Nunanuna 82
Nu 26
Nunu 58

'Ohai 67
'Ohelo kai 60
Omumu 38
'O'uru 70
Pa'anga 121
Pa'anga 'ae kuma 127
Pago 34
Pakao 50
Pakopako 130
Palaga hilitai 56
Palulu 123
Pamoko 51
Panao 31
Panopano 31
Pao 36
Papiro 51
Papati 123
Pa'u-o-Hi'iaka 126
Pepe 41
Pia 106
Pipi 123
Pipi tatahi 128
Piripiri 129
Pitipiti'o 113
Pi'ut 74
Pofatu'ao'ao 69
Pohinahina 72
Pohue 116, 125
Pohuehue 125
Pokea 99
Polo 90
Polo Tonga 104
Pomea mata'ila 113
Popolo 105
Poroporo 90
Po'ue 124, 125, 128
Po'utukava 69
Pu'a 33
Pua pilo 51
Puapua 31
Pua taukanave 27
Pu'atea 38
Pu'a vai 38
Puka 33
Puka kula 33
Puka tea 38
Puka sea 38
Pukiawe lei 113
Puko 38
Pulu tai 83
Puopua 31
Purau 32
Puteng 20

Rama 74
Rara 73
Reva 25
Roronibebe 45
Rosarosa 32

Sagali 35
Selesele 134
Selie 42
Siale tafa 49
Siale tofa 49
Sinu dina 30
Sipi 121
Sisi tai 111
Somisomi 74
Soni 50

Tafano 31
Tahinu 45
Tai'inu 45
Taino'a 117
Tainoka 117
Talamoa 50

Talatala'amoa 50
Talie 42, 43
Talisai 42
Talisai ganu 43
Tamakomako 77
Tamanu 22
Tamole 99, 100
Tatagia 19
Tatangia 19
Tataramoa 50
Tauanave 27
Tausuni 45
Tau'unu 45
Tavola 42
Tefifi 123
Telie 42
Telie 'a manu 43
Ti'anina 33
Titi tai 112
Toa 23
Toihune 45
Toihune fifine 91
To'ito'i 66
Togo 39
Tokaibebe 56
Tomitomi 74
Tongo 39, 40
Tongolei 21
Toroire 41
Toromiro 41
Toto 25, 83
Totolu 83
Totototo 83
Totoyava 83
Tou 27
Touhuni 45
Tupe 127
Tutae pua'a 127
Tutu 54
Tutu hina 53
Tutui faraoa 114
Tuvakei 107

Ufa 32
Ufi'atuli 81
Unuoi 57
'Urio 111
'Utu 20

Valai 121, 127
Vao 36
Vau 34
Vauvau 59
Vavae 59
Vavai 59, 111
Vere 53, 54
Verevere 53
Veveda 66
Vitahi 74
Volovalo 64
Vusolevu 54
Vutu 20

Wa ika 124
Wakore 127
Wa lai 121
Wa lutu mai lagi 117
Wa mila 124
Wa vere lagi 117
Wa vulavula 125
Wiliwili haole 29

Yabia 106
Yaro 64